Elementary Differential Equations

Elementary Differential Equations

Rosalind Lake

Larsen & Keller
www.larsen-keller.com

Elementary Differential Equations
Rosalind Lake
ISBN: 978-1-64172-121-9 (Hardback)

 Larsen & Keller

Published by Larsen and Keller Education,
5 Penn Plaza,
19th Floor,
New York, NY 10001, USA

Cataloging-in-Publication Data

Elementary differential equations / Rosalind Lake.
    p. cm.
Includes bibliographical references and index.
ISBN 978-1-64172-121-9
1. Differential equations. 2. Mathematics.
I. Lake, Rosalind.
QA371 .E44 2019
515.35--dc23

For more information regarding Larsen and Keller Education and its products, please visit the publisher's website www.larsen-keller.com

Table of Contents

Preface

A differential equation is a mathematical equation containing functions and their derivatives. A function represents a physical quantity, its derivative expresses the rate of change and the equation describes the relationship between the two. Differential equations can be ordinary/partial, linear/non-linear and homogeneous/inhomogeneous. These equations are also described by the order of the highest derivative into first-order, second-order differential equations and so on. These equations are important in engineering, economics, physics and biology. Some notable differential equations are the Laplace equation, the harmonic oscillator differential equation, the Euler-Lagrange equation, heat equation, Black-Scholes equation, etc. This book is a valuable compilation of topics, ranging from the basic to the most complex theories in the field of differential equations. It presents this complex subject in the most comprehensible and easy to understand language. It is an essential guide for both academicians and those who wish to pursue this discipline further.

A foreword of all Chapters of the book is provided below:

Chapter 1, A differential equation is a mathematical tool, which associates some function with its derivatives. The functions represent physical quantities, while the derivatives signify rates of change and the differential equation establishes a relationship between the two. The chapter closely examines differential equations and their various types, such as ordinary and partial differential equations, linear and non-linear differential equations, stochastic differential equation, etc.; **Chapter 2,** The order of a differential equation is determined by the order of the highest derivative, such as, in a first order differential equation the highest term has a first derivative. The power of the highest derivative of a differential equation is called its degree. The aim of this chapter is to explore the fundamentals of first order and first degree differential equations through an analysis of the crucial topics of Separation of variables Bernoulli's equation, homogeneous differential equations, exact differential equations, etc.; **Chapter 3,** A differential equation, which contains unknown multivariable functions and their partial derivatives, is called a partial differential equation. All the diverse aspects of partial differential equations such as non-linear equations, Green's function, first- and second-order partial differential equation, fundamental solution, Euler–Tricomi equation, etc. have been carefully analyzed in this chapter; **Chapter 4,** A differential equation that is subject to a set of constraints is a boundary value problem. The solution to such a differential equation satisfies the boundary conditions. In order to completely understand boundary value problems and boundary conditions, it is vital to understand the fundamental aspects of initial value problem, Sturm–Liouville problem, Dirichlet problem and elliptic boundary value problem.

I would like to thank the entire editorial team who made sincere efforts for this book and my family who supported me in my efforts of working on this book. I take this opportunity to thank all those who have been a guiding force throughout my life.

Rosalind Lake

Differential Equations and its Types

A differential equation is a mathematical tool, which associates some function with its derivatives. The functions represent physical quantities, while the derivatives signify rates of change and the differential equation establishes a relationship between the two. The chapter closely examines differential equations and their various types, such as ordinary and partial differential equations, linear and non-linear differential equations, stochastic differential equation, etc.

Differential equation is a mathematical statement containing one or more derivatives i.e. terms representing the rates of change of continuously varying quantities. Differential equations are very common in science and engineering, as well as in many other fields of quantitative study, because what can be directly observed and measured for systems undergoing changes are their rates of change. The solution of a differential equation is, in general, an equation expressing the functional dependence of one variable upon one or more others; it ordinarily contains constant terms that are not present in the original differential equation. Another way of saying this is that the solution of a differential equation produces a function that can be used to predict the behavior of the original system, at least within certain constraints.

Differential equations are classified into several broad categories, and these are in turn further divided into many subcategories. The most important categories are ordinary differential equations and partial differential equations. When the function involved in the equation depends on only a single variable, its derivatives are ordinary derivatives and the differential equation is classed as an ordinary differential equation. On the other hand, if the function depends on several independent variables, so that its derivatives are partial derivatives, the differential equation is classed as a partial differential equation.

The following are examples of ordinary differential equations:

$$\frac{dy}{dt} = -ky,$$

$$m\frac{d^2y}{dt^2} = -k^2y,$$

$$\left[1 + \left(\frac{dy}{dx}\right)^2\right]\frac{d^3y}{dx^3} - 3\frac{dy}{dx}\left(\frac{d^2y}{dx^2}\right)^2 = 0$$

In these, y stands for the function, and either t or x is the independent variable. The symbols k and m are used here to stand for specific constants.

Whichever the type may be, a differential equation is said to be of the nth order if it involves a derivative of the nth order but no derivative of an order higher than this. The equation $\frac{\partial u}{\partial t} = k^2\left[\frac{\partial^2 u}{\partial x^2} + \frac{\partial^2 u}{\partial y^2} + \frac{\partial^2 u}{\partial z^2}\right]$ is an example of a partial differential equation of the second order. The theories of ordinary and partial differential equations are markedly different, and for this reason the two categories are treated separately.

Instead of a single differential equation, the object of study may be a simultaneous system of such equations. The formulation of the laws of dynamics frequently leads to such systems. In many cases, a single differential equation of the nth order is advantageously replaceable by a system of n simultaneous equations, each of which is of the first order, so that techniques from linear algebra can be applied.

An ordinary differential equation in which, for example, the function and the independent variable are denoted by y and x is in effect an implicit summary of the essential characteristics of y as a function of x. These characteristics would presumably be more accessible to analysis if an explicit formula for y could be produced. Such a formula, or at least an equation in x and y (involving no derivatives) that is deducible from the differential equation, is called a solution of the differential equation. The process of deducing a solution from the equation by the applications of algebra and calculus is called solving or integrating the equation. It should be noted, however, that the differential equations that can be explicitly solved form but a small minority. Thus, most functions must be studied by indirect methods. Even its existence must be proved when there is no possibility of producing it for inspection. In practice, methods from numerical analysis, involving computers, are employed to obtain useful approximate solutions.

General Differential Equation

The most general differential equation in two variables is –

$$f(x, y, y', y'' \ldots\ldots) = c$$

Where,

- $f(x, y, y', y''\ldots)$ is a function of x, y, y', $y''\ldots$ and so on.
- x is the independent variable.
- y is the dependent variable.
- y', $y''\ldots$ and so on, is the first order derivative of y, second order derivative of y, and so on.
- c is some constant.

Examples of Differential Equations

Example:

It involves a derivative, $\dfrac{dy}{dx}$:

$$\frac{dy}{dx} = x^2 - 3$$

As we did before, we will integrate it. This will be a general solution (involving K, a constant of integration).

So we proceed as follows:

$$y = \int (x^2 - 3)dx$$

and this gives,

$$y = \frac{x^3}{3} - 3x + k$$

But where did that dy go from the $\frac{dy}{dx}$? Why did it seem to disappear?

In this example, we appear to be integrating the x part only (on the right), but in fact we have integrated with respect to y as well (on the left). DEs are like that - you need to integrate with respect to two (sometimes more) different variables, one at a time.

We could have written our question only using differentials:

$$dy = (x^2 - 3)dx$$

Now we integrate both sides, the left side with respect to y (that's why we use "dy") and the right side with respect to x (that's why we use "dx"):

$$\int dy = \int (x^2 - 3)dx$$

Then the answer is the same as before, but this time we have arrived at it considering the dy part more carefully:

$$y = \frac{x^3}{3} - 3x + K$$

On the left hand side, we have integrated $\int dy = \int 1 dy$ to give us y.

Note about the constant: We have integrated both sides, but there's a constant of integration on the right side only. What happened to the one on the left? The answer is quite straightforward. We do actually get a constant on both sides, but we can combine them into one constant (K) which we write on the right hand side.

Example:

This example also involves differentials:

$$\theta^2 d\theta = sin(t + 0.2)dt$$

We have:

A function of θ with $d\theta$ on the left side, and

A function of t with dt on the right side.

To solve this, we would integrate both sides, one at a time, as follows:

$$\int \theta^2 d\theta = \int \sin(t + 0.2) dt$$

$$\frac{\theta^3}{3} = -\cos(t + 0.2) + K$$

We have integrated with respect to θ on the left and with respect to t on the right.

Here is the graph of our solution, taking $K = 2$:

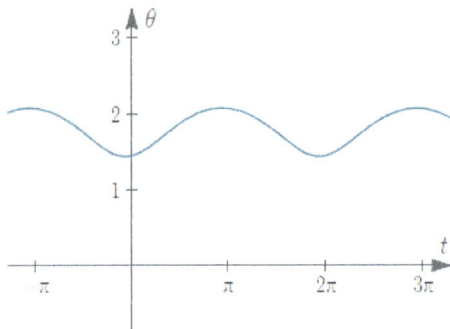

Typical solution graph for the above Example is 2 DE:

$$\theta(t) = \sqrt[3]{-3\cos(t + 0.2) + 6}$$

Solving a Differential Equation

From the above examples, we can see that solving a DE means finding an equation with no derivatives that satisfies the given DE. Solving a differential equation always involves one or more integration steps.

It is important to be able to identify the type of DE we are dealing with before we attempt to solve it.

Order and Degree

The order of a differential equation is the order of the highest order derivative involved in the differential equation. The degree of a differential equation is the exponent of the highest order derivative involved in the differential equation when the differential equation satisfies the following conditions:

- All of the derivatives in the equation are free from fractional powers, positive as well as negative if any.

- There is no involvement of the derivatives in any fraction.

- There shouldn't be involvement of highest order derivative as a transcendental function, trigonometric or exponential, etc. The coefficient of any term containing the highest order derivative should just be a function of x, y, or some lower order derivative.

If one or more of the aforementioned conditions are not satisfied by the differential equation, it should be first reduced to the form in which it satisfies all of the above conditions. An equation has no degree or undefined degree if it is not reducible.

The determination of the degree of a given differential equation can be very tricky if you are not well versed with the conditions under which the degree of the differential equation is defined.

Examples

Find the orders

For a differential equation represented by a function $f(x, y, y') = 0$; the first order derivative is the highest order derivative that has involvement in the equation. Thus, the Order of such a Differential Equation = 1. In a similar way, work out the examples below to understand the concept better –

$$x\frac{d^2y}{dx^2} + y\frac{dy}{dx} + 4y^2 = 1 : Order = 2$$

$$\sin\left(\frac{d^3y}{dx^3}\right) = \frac{dy}{dx} + x : Order = 3$$

$$For \sqrt{\left(\frac{dy}{dx}\right)^2 + 3y} = \frac{d^2y}{dx^2} : Order = 2$$

Find the Degrees

$$3y^2\left(\frac{dy}{dx}\right)^3 - \frac{d^2y}{dx^2} = sin(x^2)$$

The highest order derivative involved here is of order 2, and its power = 1 in the equation. Thus, the order of the differential equation = 2, degree = 1.

$$\sqrt{1 + \left(\frac{dy}{dx}\right)^2} = y\frac{d^3y}{dx^3}$$

Since this equation involves fractional powers, we must first get rid of them. On squaring the equation, we get $1 + \left(\frac{dy}{dx}\right)^2 = y^2\left(\frac{d^3y}{dx^3}\right)^2$ Now, we can clearly make out that the highest order derivative is of order 3 here i.e. order of the differential equation = 3 and since its power is 2 in the equation – the degree of the differential equation = 2.

$$sin\left(\frac{dy}{dx}\right) + \frac{d^2y}{dx^2} + 3x = 0$$

Here, the highest order derivative is of order 2, and it has no involvement in any function. So, the order of the differential equation = 2, and degree = 1.

More Examples

$$e^{\frac{d^2y}{dx^2}} + sin(x)\frac{dy}{dx} = 1$$

Here, the highest order derivative (order = 2) has involvement in an exponential function. Note that the exponential function can be expanded as a series to bring it to a polynomial form i.e.

$$e^x = 1 + x + \frac{x^2}{2!} + \frac{x^3}{3!}.....$$

Thus, the powers of the 2nd order derivative in the equation above will keep on varying as we incorporate more and more terms in the series expansion of the exponential function. Thus, the degree of the equation = Not Defined. The order of the equation = 2.

$$sin\left(\frac{dy}{dx}\right)\frac{d^3y}{dx^3} - \log(y) = x^2$$

Here, the coefficient of the highest order derivative is a function only of $\frac{dy}{dx}$, which is a lower order derivative. It thus defines the degree of the equation. Even if you expand the trigonometric sine function, you'll get something like $sinx = x - \frac{x^3}{3!} + \frac{x^5}{5!}.....$, which is a polynomial function with an infinite number of terms.

Since it is multiplied with $\frac{d^3y}{dx^3}$ every term in the expansion would contain the term $\frac{d\,y}{dx}$, thus maintaining the degree of the highest order derivative constant, unlike our last example. Thus, the order of the differential equation = 3 and degree = 1.

Example: $\frac{d^4y}{dx^4} + \left(\frac{d^2y}{dx^2}\right)^2 - 3\frac{dy}{dx} + y = 9$

Here, the exponent of the highest order derivative is one and the given differential equation is a polynomial equation in derivatives. Hence, the degree of this equation is 1.

Example: $\left[\frac{d^2y}{dx^2} + \left(\frac{dy}{dx}\right)^2\right]^4 = k^2\left(\frac{d^3y}{dx^3}\right)^2$

The order of this equation is 3 and the degree is 2 as the highest derivative is of order 3 and the exponent raised to highest derivative is 2.

Example: $\frac{d^2y}{dx^2} + cos\frac{d^2y}{dx^2} = 5x$

The given differential equation is not a polynomial equation in derivatives. Hence, the degree for this equation is not defined.

Example: $\left(\frac{d^3y}{dx^3}\right)^2 + y = 0$

The order of this equation is 3 and the degree is 2.

Example: Figure out the order and degree of differential equation that can be formed from the equation $\sqrt{1-x^2} + \sqrt{1-y^2} = k(x-y)$.

Solution:

Let $x = \sin\theta, y = \sin\phi$

So, the given equation can be rewritten as,

$$\sqrt{1-\sin\theta^2} + \sqrt{1-\sin\phi^2} = k\left(\sin\theta - \sin\phi\right)$$

$$\Rightarrow (\cos\theta + \cos\phi) = k(\sin\theta - \sin\phi)$$

$$\Rightarrow 2\cos\frac{\theta+\phi}{2}\cos\frac{\theta-\phi}{2} = 2k\cos\frac{\theta+\phi}{2}\sin\frac{\theta-\phi}{2}$$

$$\cot\frac{\theta-\phi}{2} = k$$

$$\theta - \phi = 2\cot^{-1}k$$

$$\sin^{-1}x - \sin^{-1}y = 2\cot^{-1}k$$

Differentiating both sides w. r. t. x, we get

$$\frac{1}{1-x^2} - \frac{1}{1-y^2} = \frac{dy}{dx} <$$

So, the degree of the differential equation is 1 and it is a first order differential equation.

Partial Differential Equations

Partial differential equation in mathematics is an equation relating a function of several variables to its partial derivatives. A partial derivative of a function of several variables expresses how fast the function changes when one of its variables is changed; the others being held constant. The partial derivative of a function is again a function, and, if f(x, y) denotes the original function of the variables x and y, the partial derivative with respect to x—i.e., when only x is allowed to vary—is typically written as $f_x(x, y)$ or $\partial f/\partial x$. The operation of finding a partial derivative can be applied to a function that is itself a partial derivative of another function to get what is called a second-order partial derivative. For example, taking the partial derivative of $f_x(x, y)$ with respect to y produces a new function $f_{xy}(x, y)$, or $\partial^2 f/\partial y \partial x$. The order and degree of partial differential equations are defined the same as for ordinary differential equations.

In general, partial differential equations are difficult to solve, but techniques have been developed for simpler classes of equations called linear, and for classes known loosely as "almost" linear, in which all derivatives of an order higher than one occur to the first power and their coefficients involve only the independent variables.

Many physically important partial differential equations are second-order and linear. For example:

$u_{xx} + u_{yy} = 0$ (two-dimensional Laplace equation)

$u_{xx} = u_t$ (one-dimensional heat equation)

$u_{xx} - u_{yy} = 0$ (one-dimensional wave equation)

The behavior of such an equation depends heavily on the coefficients a, b, and c of $au_{xx} + bu_{xy} + cu_{yy}$. They are called elliptic, parabolic, or hyperbolic equations according as $b^2 - 4ac < 0, b^2 - 4ac = 0, or\ b^2 - 4ac > 0$, respectively. Thus, the Laplace equation is elliptic, the heat equation is parabolic, and the wave equation is hyperbolic.

Ordinary Differential Equation

An ordinary differential equation (frequently called an "ODE," "diff eq," or "diffy Q") is an equality involving a function and its derivatives. An ODE of order n is an equation of the form,

$$F\left(x, y, y', ..., y^{(n)}\right) = 0,$$

Where y is a function of x, y' dy/dx is the first derivative with respect to x, and $y^{(n)} = d^n y / d x^n$ is the nth derivative with respect to x.

Nonhomogeneous ordinary differential equations can be solved if the general solution to the homogenous version is known, in which case the undetermined coefficients method or variation of parameters can be used to find the particular solution.

An ODE of order n is said to be linear if it is of the form,

$$a_n(x)y^{(n)} + a_{n-1}(x)y^{(n-1)} + ... + a_1(x)y' + a_0(x)y = Q(x).$$

A linear ODE where Q(x) = 0 is said to be homogeneous. Confusingly, an ODE of the form,

$$y' = f\left(\frac{y}{x}\right)$$

is also sometimes called "homogeneous."

In general, an nth-order ODE has n linearly independent solutions. Furthermore, any linear combination of linearly independent functions solutions is also a solution.

Simple theories exist for first-order (integrating factor) and second-order (Sturm-Liouville theory) ordinary differential equations, and arbitrary ODEs with linear constant coefficients can be solved when they are of certain factorable forms. Integral transforms such as the Laplace transform can also be used to solve classes of linear ODEs.

While there are many general techniques for analytically solving classes of ODEs, the only practical solution technique for complicated equations is to use numerical methods. The most popular of these is the Runge-Kutta method, but many others have been developed, including the collocation method and Galerkin method. A vast amount of research and huge numbers of publications have been devoted to the numerical solution of differential equations, both ordinary and partial (PDEs) as a result of their importance in fields as diverse as physics, engineering, economics, and electronics.

The solutions to an ODE satisfy existence and uniqueness properties. These can be formally established by Picard's existence theorem for certain classes of ODEs. Let a system of first-order ODE be given by

$$\frac{dx_i}{dt} = f_i(x_1,...,x_n,t),$$

for $i = 1, ..., n$ and let the functions $f_i(x_1,...,x_n,t)$, where $i = 1, ..., n$, all be defined in a domain D of the (n+1)-dimensional space of the variables $x_1, ...,x_n,t$. Let these functions be continuous in D and have continuous first partial derivatives $\partial f_i / \partial x_j$ for $i = 1, ..., n$ and $j = 1,.....,n$ in D. Let $(x_1^o,...,x_n^o)$ be in D. Then there exists a solution of $\frac{dx_i}{dt} = f_i(x_1,...,x_n,t)$ given by

$$x_1 = x_1(t),...,x_n = x_n(t)$$

for $t_0 - \delta < t < t_0 + \delta$ (where $\delta > 0$) satisfying the initial conditions

$$x_1(t_0) \;=\; x_1^o, \; ..., \; x_n(t_0) \;=\; x_n^o.$$

Furthermore, the solution is unique, so that if

$$x_1 = x_1^*(t),...,x_n = x_n^*(t)$$

is a second solution of $\frac{dx_i}{dt} = f_i(x_1,...,x_n,t)$ for $t_0 - \delta < t < t_0 + \delta$ satisfying $x_1(t_0) = x_1^o,...,x_n(t_0) = x_n^o$, then $x_i(t) \equiv x_i^*(t)$ for $t_0 - \delta < t < t_0 + \delta$. Because every nth-order ODE can be expressed as a system of n first-order ODEs, this theorem also applies to the single nth -order ODE.

The simplest Possible ODE

Let's start simpler, though. What is the simplest possible ODE? Let $x(t)$ be a function of t that satisfies the ODE:

$$\frac{dx}{dt} = 0.$$

We can ask some simple questions. What is $x(t)$? Is $x(t)$ uniquely determined from this equation? If not, what else do you need to specify?

Equation $\dfrac{dx}{dt} = 0$ just means that $x(t)$ is a constant function, $x(t) = C$. It is certainly not uniquely determined, as there is no way to specify the constant C if we only have equations for the derivatives of x. In order to uniquely determine $x(t)$, one must provide some additional data in terms of the function $x(t)$ itself.

We could for example, specify that $x(t)$ must be equal to 31 when t=11, adding the condition

$$x(11) = 31$$

Then we know C=31 and the function is $x(t) = 31$ for all t. We frequently think of the variable t as representing time and refer to a condition such as $x(11) = 31$ as an initial condition.

Let's write the initial condition more generally as

$$x(t_0) = x_0,$$

Where t_0 is some given time and x_0 is some given number. It's as though we initialize the system to be equal to the number x_0 at the time $t = t_0$. However, this "initial condition" also determines x(t) for early times. As you can see from the solution x(t)=31 for all time t, this condition specifies the state of the system for times before and after t=11.

A Slightly more Complicated ODE

Let's make things a little more complicated. Consider the equation

$$\frac{dx}{dt} = m \sin t + nt^3,$$

where m and n are just some real numbers. Equation $\dfrac{dx}{dt} = m \sin t + nt^3$, isn't much more complicated than equation $\dfrac{dx}{dt} = 0$. because the right hand side does not depend on x. It only depends on t. We are simply specifying what the derivative is in terms of t. The solution is just the antiderivative, or the integral.

Let's do the integral slightly differently this time. We'll use the definite integral from time t=a to time t=b. Using the fundamental theorem of calculus, the integral of $\dfrac{dx}{dt}$ from a to b must be

$$x(b) - x(a) = \int \frac{dx}{dt} dt$$
$$= \int (m \sin t + nt) dt$$
$$= -m \cos b + nb^4 / 4 - (-m \cos a + na^4 / 4).$$

We can write the solution in different ways. We could just replace b with an arbitrary time t,

$$x(t) = -m \cos t + nt^4 / 4 + m \cos a - na^4 / 4 + x(a).$$

This form makes it very obvious how the solution $x(t)$ would depend on an initial condition $x(t_o) = x_o$. If $x(7) = 5$, then

$$x(t) = -m\cos t + nt^4/4 + m\cos 7 - n7^4/4 + 5.$$

On the other hand, if we aren't concerned with the form of the constant, we could just write the general solution as

$$x(t) = -m\cos t + nt^4/4 + C$$

for some arbitrary constant C.

An ODE that isn't a Simple Integral

So far, the example ODEs we've seen could be solved simply by integrating. The reason they were so simple is that the equations for $\dfrac{dx}{dt}$ did not depend on the function $x(t)$ but only on the variable t. On the other hand, once the equation depends on both $\dfrac{dx}{dt}$ and $x(t)$, we have do more work to solve for the function $x(t)$.

Here's an ODE that includes $x(t)$:

$$\frac{dx}{dt} = ax(t) + b$$

Where a and b are some constants. Since the right hand side depends on x itself, we cannot simply integrate and use the fundamental theorem of calculus. To solve this ODE for $x(t)$, we'll need to do some manipulations and use the chain rule (i.e., a u-substitution).

The first thing to do is get all expressions involving x on one side of the equation. If we subtract, we won't be able to put things in the right form for the chain rule, as we'll have terms without a $\dfrac{dx}{dt}$ in them. Instead, we divide both sides of the equation by $ax(t) + b$,

$$\frac{\dfrac{dx}{dt}}{ax(t) + b} = 1.$$

Now the right hand side is a simple function of t (a constant function in this case). We can integrate both sides of the equation with respect to t,

$$\int \frac{\dfrac{dx}{dt}\,dt}{ax(t) + b} = \int 1\,dt.$$

At first glance, the left hand side might look ugly. But it is in a special form that makes it easy to integrate. It contains a $\dfrac{dx}{dt}dt$ factor, and the remaining dependence on t is only through the function x(t). If we change variables (do a u-substitution) of the form u= $x(t)$, then du= $\dfrac{dx}{dt}dt$, and we just replace the remaining appearances of $x(t)$ with u. The left hand side is then a simple integral in terms of the new variable u, which we can integrate and substitute back u = $x(t)$:

$$\int \frac{\frac{dx}{dt}dt}{ax(t)+b} = \int \frac{du}{au+b}.$$

$$= \frac{1}{a}\log|au+b|+c_1$$

$$= \frac{1}{a}\log|ax(t)+b|+c_1,$$

For some arbitrary constant C_1.

Since this expression must be equal to $\int 1\,dt = t+C_2$ for another arbitrary constant C_2, we obtain an equation for $x(t)$ and t,

$$\frac{1}{a}\log|ax(t)+b|+C_1 = t+C_2.$$

Let $C_3 = C_2 - C_1$, and then solve the equation for $x(t)$:

$$\frac{1}{a}\log|ax(t)+b| = t+C_3$$

$$|ax(t)+b| = exp(at+aC_3)$$

$$ax(t)+b = \pm exp(at+aC_3)$$

$$x(t) = \pm\frac{1}{a}exp(at+aC_3)-b/a.$$

(The notation $exp(z)$ is just another way of writing the exponential e^z.)

We can write this equation more simply by defining a new arbitrary constant $C = \pm\dfrac{1}{a}\,exp(aC_3)$. Then, the solution to our ODE can be written as

$$x(t) = Ce^{at}-b/a$$

Can you verify that this solution for $x(t)$ does indeed satisfy the original ODE $\dfrac{dx}{dt} = ax+b$? Since checking that a solution satisfies an ODE is much easier and less error-prone than solving the ODE, verifying the solution is an essential step in the solution process.

Let's check our solution. If $x = Ce^{at}-b/a$, then $\dfrac{dx}{dt} = Cae^{at}$. On the other hand, $ax+b = Cae^{at}-b+b = Cae^{at}$. Yes, these expressions match, $\dfrac{dx}{dt} = ax+b$, and we can be confident of our solution.

In order to determine the constant C, we need an additional condition. For example, if $x(3) = 4$, then C must satisfy

$$4 = Ce^{3a} - b/a$$

so that

$$C = (4 + b/a)e^{-3a}$$

Our solution for this initial condition is

$$x(t) = (4 + b/a)e^{-3a}e^{at} - b/a$$

or

$$x(t) = (4 + b/a)e^{a(t-3)} - b/a.$$

A Shortcut method to Solving Simple ODEs

For the above solution, we did some extra steps in order to demonstrate that the manipulations were really nothing more than a u-substitution. Usually, we'll skip many of these steps and use a shortcut method. However, before jumping into the shortcut method, make sure you understand how the above u-substitution works.

Let's revisit our solution method to see how we can take some shortcuts. The first thing we could do differently is avoid changing to the variable u. We could keep everything in terms of x, in which case, the u-substitution would be replacing $x(t)$ with x and $\frac{dx}{dt} dt$ with dx.

Next, observe the results of the substitution. We started with

$$\frac{\frac{dx}{dt}}{ax + b} = 1$$

and ended up with

$$\int \frac{dx}{ax + b} = \int 1\, dt ,$$

where now we wrote everything in terms of x rather than u. To accomplish this manipulation, we multiplied by dt and did our substitution to replace $\frac{dx}{dt} dt$ by dx. It was as though we canceled the dt from the numerator with the dt from the denominator. The derivative $\frac{dx}{dt}$ isn't really a fraction of numbers dx and dt, but in an integral, applying the chain rule (i.e., u-substitution) makes it behave like it is a fraction.

Hence, in practice, we can safely treat $\frac{dx}{dt}$ like a fraction when used in this context of forming an integral to solve a differential equation. To solve the equation $\dfrac{dx}{dt} = ax + b$, we multiply both sides of the equation by dt and divide both sides of the equation by $ax + b$ to get

$$\frac{dx}{ax + b} = dt.$$

Then, we integrate both sides to obtain

$$\int \frac{dx}{ax + b} = \int dt \,,$$

Just remember that these manipulations are really a shortcut way to denote using the chain rule.

Linear Differential Equations

Linear Differential equations in mathematics refer to the differential equations in only a single variable which can be solved easily rather than having two variables in the equation. Linear Differential equations satisfy, what we may call as superposition principle, which allows us to add the solution in two different linear differential equation to get the solution of a whole new equation. This implies that a complex differential equation, if possible, can be disintegrated into linear differential equations, which makes the solving easy.

In this case, the differential equation looks like

$$a_n \frac{d^n y}{dx^n} + a_{n-1} \frac{d^{n-1} y}{dx^{n-1}} + \cdots + a_1 \frac{dy}{dx} + a_0 y = 0 \,,$$

with $a_0, a_1, a_{n-1}, \ldots, a_n$ being real constants, and almost resembles a polynomial. In fact, looking at the roots of this associated polynomial gives solutions to the differential equation.

Homogeneous Equation with Constant Coefficients

A homogeneous linear differential equation has *constant coefficients* if it has the form

$$a_0 y + a_1 y' + a_2 y'' + \cdots + a_n y^{(n)} = 0$$

where $a_1, \ldots, a_n$ are (real or complex) numbers. In other words, it has constant coefficients if it is defined by a linear operator with constant coefficients.

The study of these differential equations with constant coefficients dates back to Leonhard Euler, who introduced the exponential function e^x, which is the unique solution of the equation $f' = f$ such that $e^0 = 1$. It follows that the nth derivative of e^{cx} is $c^n e^{cx}$, and this allows solving homogeneous linear differential equations rather easily.

Let,

$$a_0 y + a_1 y' + a_2 y'' + \cdots + a_n y^{(n)} = 0$$

be a homogeneous linear differential equation with constant coefficients (that is $a_0, \ldots, a_n$ are real or complex numbers).

Searching solutions of this equation that have the form $e^{\alpha x}$ is equivalent to searching the constants α such that

$$a_0 e^{\alpha x} + a_1 \alpha e^{\alpha x} + a_2 \alpha^2 e^{\alpha x} + \cdots + a_n \alpha^n e^{\alpha x} = 0.$$

Factoring out $e^{\alpha x}$ (which is never zero) shows that α must be a root of the characteristic polynomial of the differential equation.

$$a_0 + a_1 t + a_2 t^2 + \cdots + a_n t^n$$

When these roots are all distinct, one has n distinct solutions that are not necessarily real, even if the coefficients of the equation are real. These solutions can be shown to be linearly independent, by considering the Vandermonde determinant of the values of these solutions at $x = 0, \ldots, n-1$. Together they form a basis of the vector space of solutions of the differential equation (that is, the kernel of the differential operator).

In the case where the characteristic polynomial has only simple roots, the preceding provides a complete basis of the solutions vector space. In the case of multiple roots, more linearly independent solutions are needed for having a basis. These have the form

$$x^k e^{\alpha x},$$

where k is a nonnegative integer, α is a root of the characteristic polynomial of multiplicity m, and $k < m$. For proving that these functions are solutions, one may remark that if α is a root of the characteristic polynomial of multiplicity m, the characteristic polynomial may be factored as $P(t)(t-\alpha)^m$. Thus, applying the differential operator of the equation is equivalent with applying first m times the operator $\dfrac{d}{dx} - \alpha$, and then the operator that has P as characteristic polynomial. As

$$\left(\frac{d}{dx} - \alpha \right) \left(x^k e^{\alpha x} \right) = k x^{k-1} e^{\alpha x},$$

one gets zero after $k + 1$ application of $\dfrac{d}{dx} - \alpha$.

Example

$y''' - 2y''' + 2y'' - 2y' + y = 0$ has the characteristic equation $z^4 - 2z^3 + 2z^2 - 2z + 1 = 0$.

This has zeros, i, −i, and 1 (multiplicity 2). The solution basis is thus $e^{ix}, e^{-ix}, e^x, x e^x$.

A real basis of solution is thus $\cos x, \sin x, e^x, x e^x$.

As, by the fundamental theorem of algebra, the sum of the multiplicities of the roots of a polynomial equals the degree of the polynomial, the number of above solutions equals the order of the differential equation, and these solutions form a base of the vector space of the solutions.

In the common case where the coefficients of the equation are real, it is generally more convenient to have a basis of the solutions consisting of real-valued functions. Such a basis may be obtained from the preceding basis by remarking that, if $a+ib$ is a root of the characteristic polynomial, then $a-ib$ is also a root, of the same multiplicity. Thus a real basis is obtained by using Euler's formula, and replacing $x^k e^{(a+ib)x}$ and $x^k e^{(a-ib)x}$ by $x^k e^{ax} \cos(bx)$ and $x^k e^{ax} \sin(bx)$.

Second-order Case

A homogeneous linear differential equation of the second order may be written

$$y'' + ay' + by = 0,$$

and its characteristic polynomial is

$$r^2 + ar + b.$$

If a and b are real, there are three cases for the solutions, depending on the discriminant $D = a^2 - 4b$. In all three cases, the general solution depends on two arbitrary constants c_1 and c_2.

- If $D > 0$, the characteristic polynomial has two distinct real roots α, and β. In this case, the general solution is $c_1 e^{\alpha x} + c_2 e^{\beta x}$.

- If $D = 0$, the characteristic polynomial has a double root $-a/2$, and the general solution is $(c_1 + c_2 x)e^{-ax/2}$.

- If $D < 0$, the characteristic polynomial has two complex conjugate roots $\alpha \pm \beta i$, and the general solution is $c_1 e^{(\alpha+\beta i)x} + c_2 e^{(\alpha-\beta i)x}$, which may be rewritten in real terms, using Euler's formula as $e^{\alpha x}(c_1 \cos(\beta x) + c_2 \sin(\beta x))$.

Finding the solution $y(x)$ satisfying $y(0) = d_1$ and $y'(0) = d_2$, one equates the values of the above general solution at 0 and its derivative there to d_1 and d_2, respectively. This results in a linear system of two linear equations in the two unknowns c_1 and c_2. Solving this system gives the solution for a so called Cauchy problem, in which the values at 0 for the solution of the DEQ and its derivative are specified.

Nonhomogeneous Equation with Constant Coefficients

A nonhomogeneous equation of order n with constant coefficients may be written

$$y^{(n)}(x) + a_1 y^{(n-1)}(x) + \cdots + a_{n-1} y'(x) + a_n y(x) = f(x),$$

Where $a_1, \ldots, a_n$ are real or complex numbers, f is a given function of x, and y is the unknown function (for sake of simplicity, "(x)" will be omitted in the following).

There are several methods for solving such an equation. The best method depends of the nature of the function f that makes the equation nonhomogeneous. If f is a linear combination of exponential and sinusoidal functions, then exponential response formula may be used. If, more generally, f is linear combination of functions of the form $x^n e^{ax}$, $x^n \cos ax$ and $x^n \sin ax$, where

n is a nonnegative integer and a a constant (which need not to be the same in each term), then the method of undetermined coefficients may be used. Still more general, the annihilator method applies when f satisfies a homogeneous linear differential equation, typically, a holonomic function.

The most general method is the variation of constants, which is presented here.

The general solution of the associated homogeneous equation

$$y^{(n)} + a_1 y^{(n-1)} + \cdots + a_{n-1} y' + a_n y =$$

is

$$y = u_1 y_1 + \cdots + u_n y_n,$$

Where $(y_1, \ldots, y_n)$ is a basis of the vector space of the solutions and $u_1, \ldots, u_n$ are arbitrary constants. The method of variation of constants takes its name that, instead of considering $u_1, \ldots, u_n$ as constants, they are considered as functions that have to be determined for making y a solution of the nonhomogeneous equation. For this purpose, one adds the constraints

$$0 = u_1' y_1 + u_2' y_2 + \cdots + u_n' y_n$$

$$0 = u_1' y_1' + u_2' y_2' + \cdots + u_n' y_n'$$

$$\ldots$$

$$0 = u_1' y_1^{(n-2)} + u_2' y_2^{(n-2)} + \cdots + u_n' y_n^{(n-2)},$$

which imply (by product rule and induction)

$$y^{(i)} = u_1 y_1^{(i)} + \cdots + u_n y_n^{(i)}$$

for $i = 1, \ldots, n-1$, and

$$y^{(n)} = u_1 y_1^{(n)} + \cdots + u_n y_n^{(n)} + u_1' y_1^{(n-1)} + u_2' y_2^{(n-1)} + \cdots + u_n' y_n^{(n-1)}.$$

Replacing in the original equation y and its derivative by these expression, and using the fact that $y_1, \ldots, y_n$ are solutions of the original homogeneous equation, one gets

$$f = u_1' y_1^{(n-1)} + \cdots + u_n' y_n^{(n-1)}.$$

One has thus a system of n linear equations in $u_1', \ldots, u_n'$, which can be solved by any method of linear algebra. Then the computation of antiderivatives gives $u_1, \ldots, u_n$ and then $y = u_1 y_1 + \cdots + u_n y_n$.

As antiderivatives are defined up to the addition of a constant, one finds again that the general solution of the nonhomogeneous equation is the sum of an arbitrary solution and the general solution of the associated homogeneous equation.

Non-linear Differential Equations

A non-linear differential equation is simply a differential equation where some non-linearity is applied to either the inputs or the outputs of the equation. There are two types of non-linearity which can be encountered: functional non-linearities; and combinational non-linearities.

Functional Non-linearities

These are the easiest to spot - if an non-linear function is applied to any of the inputs or outputs, then the differential equation is non-linear. Some examples to non-linear differential equations with functional non-linearities are given below. In all the examples 'y' is the output and 'u' is the input (both y and u are functions of time).

$$\tau \frac{dy}{dt} + \sqrt{y} = u$$

$$\tau^2 \frac{d^2y}{dt^2} + \alpha \left(\frac{dy}{dt} \right)^2 + y = Ku$$

$$\tau \frac{dy}{dt} + y = \sin(u)$$

Combinational Non-linearities

Combinational non-linearities are more difficult to spot and arise when two time-varying quantities are multiplied, or divided, together. These sort of non-linearities can arise, or disappear, depending on the assumptions made about the variables in an equation. For example, the equation below:

$$\tau \frac{dy}{dt} + y = Ku_1 u_2$$

Is linear if either input 'u_1' or 'u_2' is assumed constant for the solution, but is non-linear if both are assumed to be variable. It is generally a good idea when solving things analytically to assume that everything is constant other than the things that you definitely know aren't.

First Order Differential Equation

A first-order differential equation is an equation,

$$\frac{dy}{dx} = f(x,y)$$

in which $f(x,y)$ is a function of two variables defined on a region in the xy-plane. The equation is of first order because it involves only the first derivative dy/dx (and not higher-order derivatives). We point out that the equations,

$$y' = f(x,y) \ and \ \frac{d}{dx} y = f(x,y)$$

are equivalent to Equation $\frac{dy}{dx} = f(x,y)$ and all three forms will be used interchangeably in the text. A solution of Equation $\frac{dy}{dx} = f(x,y)$ is a differentiable function defined on an interval I of x-values (perhaps infinite) such that

$$\frac{d}{dx} y(x) = f(x, y(x))$$

on that interval. That is, when y(x) and its derivative $y'(x)$ are substituted into Equation $\frac{dy}{dx} = f(x,y)$, the resulting equation is true for all x over the interval I. The general solution to a first order differential equation is a solution that contains all possible solutions. The general solution always contains an arbitrary constant, but having this property doesn't mean a solution is the general solution. That is, a solution may contain an arbitrary constant without being the general solution.

Example: Show that every member of the family of functions

$$y = \frac{C}{x} + 2$$

is a solution of the first-order differential equation

$$\frac{dy}{dx} = \frac{1}{x}(2 - y)$$

on the interval $(0, \infty)$, where C is any constant

Solution: Differentiating $y = C / x + 2$ gives

$$\frac{dy}{dx} = C \frac{d}{dx}\left(\frac{1}{x}\right) + 0 = -\frac{C}{x^2}$$

Thus we need only verify that for all $x \in (0, \infty)$

$$-\frac{C}{x^2} = \frac{1}{x}\left[2 - \left(\frac{C}{x} + 2\right)\right]$$

This last equation follows immediately by expanding the expression on the right-hand side:

$$\frac{1}{x}\left[2 - \left(\frac{C}{x} + 2\right)\right] = \frac{1}{x}\left(-\frac{C}{x}\right) = -\frac{C}{x^2}$$

Therefore, for every value of C, the function $y = \frac{C}{x} + 2$ is a solution of the differential equation.

As was the case in finding ant derivatives, we often need a particular rather than the general solution to a first-order differential equation $y' = f(x, y)$. The particular solution satisfying the initial condition $y(x_0) = y_0$ is the solution $y = y(x)$ whose value is y_0 when $x = x_0$. Thus the graph of the particular solution passes through the point (x_0, y_0) in the xy-plane. A first-order initial value problem is a differential equation $y' = f(x, y)$ whose solution must satisfy an initial condition $y(x_0) = y_0$.

Example: Show that the function

$$y = (x+1) - \frac{1}{3} e^x$$

is a solution to the first-order initial value problem

$$\frac{dy}{dx} = y - x, \qquad y(0) = \frac{2}{3}$$

Solution: The equation

$$\frac{dy}{dx} = y - x$$

is a first-order differential equation with $f(x, y) = y - x$.

On the left side of the equation:

$$\frac{dy}{dx} = \frac{d}{dx}\left(x + 1 - \frac{1}{3} e^x\right) = 1 - \frac{1}{3} e^x.$$

On the right side of the equation:

$$y - x = (x+1) - \frac{1}{3} e^x - x = 1 - \frac{1}{3} e^x$$

The function satisfies the initial condition because

$$y(0) = \left[(x+1) - \frac{1}{3} e^x\right]_{x=0} = 1 - \frac{1}{3} = \frac{2}{3}.$$

The graph of the function is shown in below Figure.

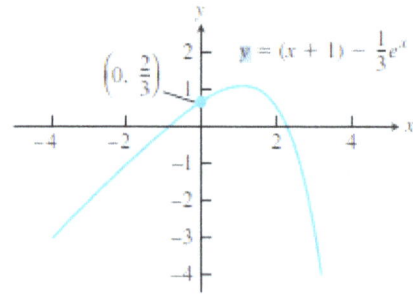

Figure : Graph of the solution $y = (x+1) - \frac{1}{3} e^x$ *to the differential equation* $\frac{dy}{dx} = y - x$, *with initial condition* $y(0) = \frac{2}{3}$

Slope Fields: Viewing Solution Curves

Each time we specify an initial condition $y(x_o) = y_o$ for the solution of a differential equation $y' = f(x,y)$, the solution curve (graph of the solution) is required to pass through the point (x_o, y_o) and to have slope $f(x_o, y_o)$ there. We can picture these slopes graphically by drawing short line segments of slope $f(x,y)$ at selected points (x,y) in the region of the xy-plane that constitutes the domain of f. Each segment has the same slope as the solution curve through (x,y) and so is tangent to the curve there. The resulting picture is called a slope field (or direction field) and gives a visualization of the general shape of the solution curves. In below Figure a shows a slope field, with a particular solution sketched into it in below Figure. We see how these line segments indicate the direction the solution curve takes at each point it passes through.

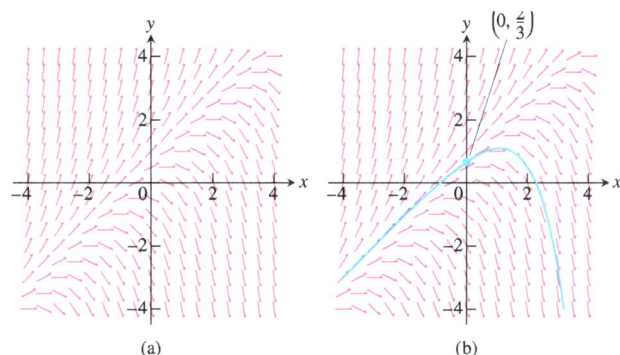

(a)   (b)

Figure: (a) Slope field for $\dfrac{dy}{dx} = y - x$. (b) The particular solution curve through the point $\left(0, \dfrac{2}{3}\right)$

Below Figure shows three slope fields and we see how the solution curves behave by following the tangent line segments in these fields.

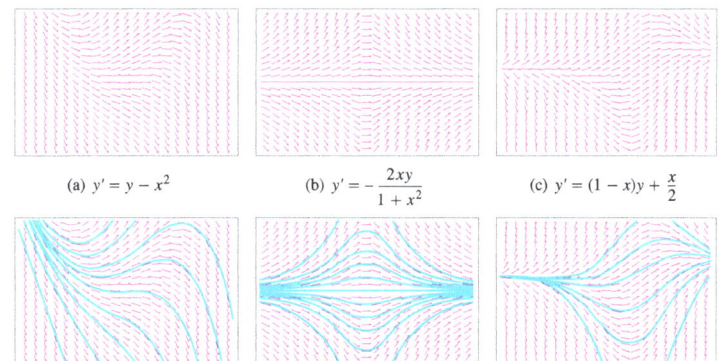

(a) $y' = y - x^2$ (b) $y' = -\dfrac{2xy}{1 + x^2}$ (c) $y' = (1 - x)y + \dfrac{x}{2}$

Figure: Slope fields (top row) and selected solution curves (bottom row). In computer renditions, slope segments are sometimes portrayed with arrows, as they are here. This is not to be taken as an indication that slopes have directions, however, for they do not.

Constructing a slope field with pencil and paper can be quite tedious. All our examples were generated by a computer.

The Existence of Solutions

A basic question in the study of first-order initial value problems concerns whether a solution even exists. A second important question asks whether there can be more than one solution. Some conditions must be imposed to assure the existence of exactly one solution, as illustrated in the next example.

Example: The initial value problem

$$\frac{dy}{dx} = y^{4/5}, \qquad y(0) = 0$$

Has more than one solution. One solution is the constant function $y(x) = 0$ for which the graph lies along the x-axis. A second solution is found by separating variables and integrating. This leads to

$$y = \left(\frac{x}{5}\right)^5$$

The two solutions $y = 0$ and $y = (x/5)^5$ both satisfy the initial condition $y(0) = 0$

We have found a differential equation with multiple solutions satisfying the same initial condition. This differential equation has even more solutions. For instance, two additional solutions are

$$y = \begin{cases} 0, & \text{for } x \leq 0 \\ \left(\frac{x}{5}\right)^5, & \text{f or } x > 0 \end{cases}$$

and

$$y = \begin{cases} \left(\frac{x}{5}\right)^5, & \text{for } x \leq 0 \\ 0, & \text{f or } x > 0 \end{cases}$$

Second Order Differential Equation

A second order differential equation is written in general form as

$$F(x, y, y', y'') = 0,$$

Where F is a function of the given arguments.

If the differential equation can be resolved for the second derivative y'', it can be represented in the following explicit form:

$$y'' = f(x, y, y').$$

In special cases the function f in the right side may contain only one or two variables. Such in-complete equations include 5 different types:

$$y'' = f(x), \ y'' = f(y), \ y'' = f(y'), \quad y'' = f(x,y'), \quad y'' = f(y,y').$$

With the help of certain substitutions, these equations can be transformed into first order equations.

In the general case of a second order differential equation, its order can be reduced if this equation has a certain symmetry. Below we discuss two types of such equations (cases 6 and 7):

- The function $F(x,y,y',y'')$ is a homogeneous function of the arguments y,y',y''

- The function $F(x,y,y',y'')$ is an exact derivative of the first order function $\Phi(x,y,y')$.

Consider these cases of reduction of order in more detail.

Case 1. Equation of type $y'' = f(x)$

For an equation of type $y'' = f(x)$ its order can be reduced by introducing a new function $p(x)$ such that $y' = p(x)$. As a result, we obtain the first order differential equation $p' = f(x)$

Solving it, we find the function $p(x)$. Then we solve the second equation

$$y' = p(x)$$

and obtain the general solution of the original equation.

Case 2. Equation of type $y'' = f(y)$

The right-hand side of the equation depends only on the variable y. We introduce a new function $p(y)$, setting $y' = p(y)$. Then we can write:

$$y'' = \frac{d}{dx}(y') = \frac{dp}{dx} = \frac{dp}{dy}\frac{dy}{dx} = \frac{dp}{dy}p,$$

so the equation becomes:

$$\frac{dp}{dy}p = f(y).$$

Solving it, we find the function $p(y)$. Then we find the solution of the equation $y' = p(y)$, that is, the function $y(x)$.

Case 3. Equation of type $y'' = f(y')$

In this case, to reduce the order we introduce the function $y' = p(x)$ and obtain the equation $y'' = p' = \dfrac{dp}{dx} = f(p)$, which is a first order equation with separable variables p and x. Integrating, we find the function $p(x)$, and then the function $y(x)$.

Case 4. Equation of type $y'' = f(x,y')$

Here we use the substitution $y' = p(x)$, where $p(x)$ is a new unknown function. As a result, we obtain the first order equation:

$$p' = \frac{dp}{dx} = f(x,p).$$

By integrating, we find the function $p(x)$. Next, we solve one more equation of the 1st order,

$$y' = p(x)$$

and find the general solution $y(x)$.

Case 5. Equation of type $y'' = f(y,y')$

To solve this equation, we introduce a new function $p(y)$, setting $y' = p(y)$, similar to case 2. Differentiating this expression with respect to x leads to the equation,

$$y'' = \frac{d(y')}{dx} = \frac{dp}{dx} = \frac{dp}{dy}\frac{dy}{dx} = \frac{dp}{dy}p.$$

As a result, our original equation is written as an equation of the 1st order,

$$p\frac{dp}{dy} = f(y,p).$$

Solving it, we find the function $p(y)$. Then we solve another first order equation,

$$y' = p(y)$$

and determine the general solution $y(x)$.

The above 5 cases of reduction of order are not independent. Based on the structure of the equations, it is clear that case 2 follows from the case 5 and case 3 follows from the more general case 4.

Case 6. Function $F(x,y,y',y'')$ is homogeneous with respect to the arguments y,y',y''

If the left side of the differential equation,

$$F(x,y,y',y'') = 0$$

satisfies the condition of homogeneity, i.e. the relationship,

$$F(x,ky,ky',ky'') = k^m F(x,y,y',y'')$$

is valid for any k, the order of the equation can be reduced by substitution,

$$y = e^{\int z dx}.$$

After the function $z(x)$ is found, the original function $y(x)$ is determined by the integration formula,

$$y(x) = C_2 e^{\int z dx},$$

where C_2 is the constant of integration.

Case 7. Function $F(x,y,y',y'')$ is an exact derivative.

If one can find a function $\Phi(x,y,y')$, which does not contain the second derivative y'' and satisfies the equation,

$$F(x,y,y',y'') = \frac{d}{dx}\Phi(x,y,y'),$$

then the solution of the original equation is given by the integral,

$$\Phi(x,y,y') = C.$$

Using this way the second order equation can be reduced to first order equation.

Delay Differential Equation

Delay-differential equations (DDEs) are a large and important class of dynamical systems. They often arise in either natural or technological control problems. In these systems, a controller monitors the state of the system, and makes adjustments to the system based on its observations. Since these adjustments can never be made instantaneously, a delay arises between the observation and the control action.

There are different kinds of delay-differential equations. We will focus on just one kind, namely those of the form

$$\dot{x} = f\left(x(t), x(t-\tau_1), x(t-\tau_2), \ldots, x(t-\tau_n)\right),$$

where the quantities τ_i are positive constants. In other words, we will focus on equations with fixed, discrete delays. There are other possibilities, notably equations with state-dependent delays (the τ_i's depend on x) or with distributed delays (the right-hand side of the differential equation is a weighted integral over past states).

When we give initial conditions for finite-dimensional dynamical systems, we only need to specify a small set of numbers, namely the initial values of the state variables, and perhaps the initial time in nonautonomous systems. In order to solve a delay equation, we need more: At every time step, we have to look back to earlier values of x. We therefore need to specify an initial function which gives the behavior of the system prior to time 0 (assuming that we start at t = 0). This function has to cover a period at least as long as the longest delay since we will be looking back in time that far.

Let us for the moment specialize further to equations with a single delay, i.e

$$\dot{x} = f\left(x(t), x(t-\tau)\right)$$

The initial function would be a function $x(t)$ defined on the interval $[-\tau,0]$. How are we to understand the dynamics induced by this delay equation? We could think in the same terms as we do for ordinary differential equations, namely that the solution consists of a sequence of values of $\mathbf{x}$ at increasing values of t. From a purely theoretical perspective however, this is not the best way to think of equations of this type. A much better way is to think of the solution of this DDE as a mapping from functions on the interval $[t-\tau,t]$ into functions on the interval $[t,t+\tau]$. In other words, the solutions of this dynamical system can be thought of as a sequence of functions $f_0(t), f_1(t), f_2(t),...$, defined over a set of contiguous time intervals of length τ. The points $t = 0, \tau, 2\tau,...$ where the solution segments meet are called knots. Given below figure illustrates this mapping. Existence and uniqueness theorems analogous to those for ordinary differential equations are much more easily proven in this conceptual framework than by trying to think of DDEs as an evolution over the state space $\mathbf{x}$.

In some very simple cases, we can work out this mapping analytically, as the following example demonstrates.

Example: Consider the delay-differential equation

$$\dot{x} = -x(t-1)$$

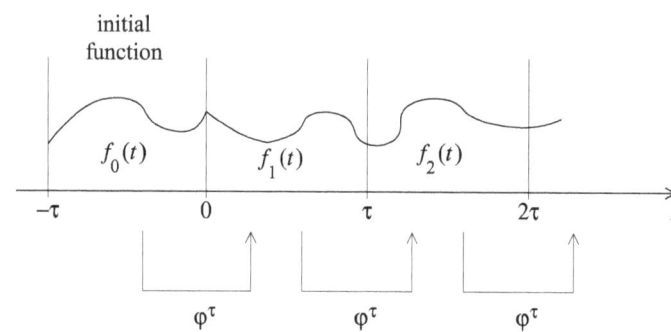

Figure: The action of the evolution operator ϕ^τ for a delay-differential equation of the form 2 is to take a function defined over a time interval of length τ and to map it into another function defined over a similar time interval. For instance, the initial function defined over the interval $[-\tau, 0]$ is mapped into a solution curve on the interval $[0, \tau]$.

Suppose that we have $x(t) = f_{i-1}(t)$ over some interval $[t_i - 1, t_i]$. then, over the interval $[t_i, t_i + 1]$, we have, by separation of variables,

$$\int_{f_{i-1}(t_i)}^{x(t)} dx' = -\int_{t_i}^{t} f_{i-1}(t'-1)dt'$$

$$\therefore x(t) = f_i(t) = f_{i-1}(t_i) - \int_{t_i}^{t} f_{i-1}(t'-1)\, dt'.$$

The foregoing example provides the basic theory for the simplest method for solving DDEs, known as the method of steps. The following example demonstrates its use to obtain a solution given some initial data.

Example: Suppose that we are given the problem of solving equation 3 given the constant initial data

$$x(t) = 1 \qquad \text{for} \quad t \in [-1, 0].$$

In the interval $[0,1]$, we have

$$x(t) = 1 - \int_0^t 1 \, dt' = 1 - t.$$

In the interval $[1,2]$, we have

$$x(t) = 0 - \int_1^t \left[1 - (t'-1)\right] dt' = -\left[2t - \frac{1}{2}t^2\right]_1^t = -2t + \frac{1}{2}t^2 + \frac{3}{2}.$$

On $[2,3]$, the solution is

$$x(t) = -\frac{1}{2} - \int_2^t \left[-2(t'-1) + \frac{1}{2}(t'-1)^2 + \frac{3}{2}\right] dt'$$

$$= -\frac{1}{2} - \left[-(t'-1)^2 + \frac{1}{6}(t'-1)^3 + \frac{3}{2}t'\right]_2^t$$

$$= \frac{5}{3} + (t-1)^2 - \frac{1}{6}(t-1)^3 - \frac{3}{2}t.$$

We can of course automate these calculations fairly easily in Maple. The following Maple procedure computes the f_i's defined in the previous example, f_o being the initial function:

> f := proc(i,t) option remember;

> if i=0 then 1

> else f(i-1,i-1) - int(f(i-1,xi-1),xi=i-1..t);

> fi; end;

This routine depends on the fact that the delay is 1 so that the ends of the intervals over which each of the f_i's are valid are just $t = 0,1,2,...$ To plot the result, we use the following Maple commands:

> with(plots):

> for i from 0 to 20 do

> p[i] := plot(f(i,t),t=i-1..i): od:

> display([p[j] $ j=0..20]);

The result is shown in below Figure the solution is a damped oscillation made up in piecewise fashion by a set of polynomials, the functions f_i .

If you look at the solution of the simple DDE plotted in below Figure, you may notice that the first derivative of x(t) isn't continuous at the first knot, $t = 0$. This isn't surprising: For $t < 0$, we have an arbitrary initial function. For $t > 0$, the solution is dictated by the differential equation. Since we made no special attempt to match up the initial function to the solution, the derivative given by equation $\dot{x} = -x(t-1)$ was bound not to match up with the zero derivative of our initial function.

It is of course possible to cook up an initial function whose first derivative at t = 0 matches that of the solution ($f_0(t) = at$ works in this case, for any value of a), but then the second derivative won't. If you somehow manage to match up the second derivative, some higher derivative will be discontinuous at t = 0 except in some highly artificial circumstances in which you can actually find a particular solution to the DDE. In these cases, you can use this solution as the initial function, and of course all will be smooth and continuous. In general however, the derivatives of x(t) at t = 0 will be discontinuous. These discontinuities propagate: If the first derivative is discontinuous at t = 0, then the second derivative will be discontinuous at $t = \tau$ since $\ddot{x}(t)$ is related by the DDE to $\dot{x}(t-\tau)$. For instance, for equation 3, $\ddot{x}(t) = -\dot{x}(t-\tau)$ so a discontinuity in the first derivative at t = 0 becomes a discontinuity in the second derivative at $t = \tau$, then a discontinuity in the third derivative at t = 2τ, and so on.

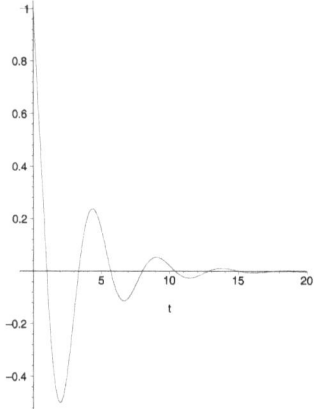

Figure: Solution of the DDE $\dot{x} = -x(t-\tau)$ obtained in Maple by the method of steps.

Most DDEs don't have analytic solutions, so it is generally necessary to resort to numerical methods. Because the solutions have discontinuous derivatives at the knots, it is necessary to be a little careful when using numerical methods designed for ODEs with DDEs.

Reduction to ODE

In some cases, differential equation can be represented in a format that looks like a delay differential equations.

Example: Consider an equation

$$\frac{d}{dt}x(t) = f\left(t, x(t), \int_{-\infty}^{0} x(t+\tau)e^{\lambda\tau}d\tau\right).$$

Introduce $y(t) = \int_{-\infty}^{0} x(t+\tau)e^{\lambda\tau}d\tau$ to get a system of ODEs

$$\frac{d}{dt}x(t) = f(t,x,y), \quad \frac{d}{dt}y(t) = x - \lambda y.$$

Example: An equation

$$\frac{d}{dt}x(t) = f\left(t, x(t), \int_{-\infty}^{0} x(t+\tau)\cos(\alpha\tau+\beta)d\tau\right)$$

is equivalent to

$$\frac{d}{dt}x(t) = f(t,x,y), \quad \frac{d}{dt}y(t) = \cos(\beta)x + \alpha z, \quad \frac{d}{dt}z(t) = \sin(\beta)x - \alpha y,$$

Where

$$y = \int_{-\infty}^{0} x(t+\tau)\cos(\alpha\tau + \beta)d\tau, \quad z = \int_{-\infty}^{0} x(t+\tau)\sin(\alpha\tau + \beta)d\tau.$$

Differential Algebraic Equations

A differential-algebraic equation (DAE) is an equation involving an unknown function and its derivatives. A (first order) DAE in its most general form is given by

$$F(t,x,x') = 0, \qquad x_i$$

Where $x = x(t)$, the unknown function, and $F = F(t,u,v)$ have N components, denoted by x_i and $F_i, i = 1,2,...,N$, respectively. Every DAE can be written as a first order DAE. The term DAE is usually reserved for the case when the highest derivative x' cannot be solved for in terms of the other terms t,x, when $F(t,x,x') = 0$, $t_0 \leq t \leq t_f$, is viewed as an algebraic relationship between three variables t,x,x' . The Jacobian $\partial F / \partial v$ along a particular solution of the DAE may be singular. Systems of equations like $F(t,x,x') = 0$, $t_0 \leq t \leq t_f$, are also called implicit systems, generalized systems, or descriptor systems. The DAE may be an initial value problem where x is specified at the initial time, $x(t_0) = x_0$, or a boundary value problem, where the solution is subject to N two-point boundary conditions $g(x(t_0),x(t_f)) = 0$.

The method of solution of a DAE will depend on its structure. A special but important class of DAEs of the form $F(t,x,x') = 0$, $t_0 \leq t \leq t_f$, is the semi-explicit DAE or ordinary differential equation (ODE) with constraints

$$y' = f(t,y,z)$$
$$0 = g(t,y,z)$$

Which appear frequently in applications. Here x=(y,z) and g(t,y,z)=0 are the explicit constraints.

Where do DAEs arise?

DAEs in either the general form $F(t,x,x') = 0$, $t_0 \leq t \leq t_f$, or the special form $\begin{matrix} y' & = & f(t,y,z) \\ 0 & = & g(t,y,z) \end{matrix}$ arise in the mathematical modeling of a wide variety of problems from engineering and science such as in multibody and flexible body mechanics, electrical circuit design, optimal control, incompressible fluids, molecular dynamics, chemical kinetics (quasi steady state and partial equilibrium approximations), and chemical process control.

Example: A simple example of a DAE arises from modeling the motion of a pendulum in Cartesian coordinates.

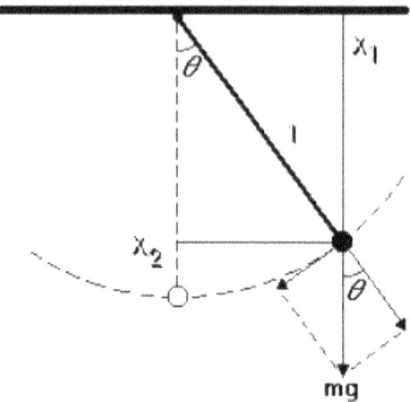

Figure: Pendulum

Suppose the pendulum has length 1 and let the coordinates of the tiny ball of mass 1 at the end of the rod be (x_1, x_2). Newton's equations of motion give,

$$x_1'' = -\lambda x_1$$
$$x_2'' = -\lambda x_2 - g,$$

Where g is the force of gravity and λ is a Lagrange multiplier. The λx_i terms represent the force which holds the solution onto the constraint,

$$x_1^2 + x_2^2 = 1,$$

which expresses the condition that the rod has fixed length 1. After rewriting the two second order equations as four first order ODEs, a DAE system of the form $\begin{aligned} y' &= f(t,y,z) \\ 0 &= g(t,y,z) \end{aligned}$ with four differential and one algebraic equations results:

$$x_1' = x_3$$
$$x_2' = x_4$$
$$x_3' = -\lambda x_1$$
$$x_4' = -\lambda x_2 - g$$
$$x_1^2 + x_2^2 = 1.$$

In this very simple case of a multibody mechanical system, the change of variables $x_1 = sin\theta$, $x_2 = cos\theta$ followed by some algebra gives the well-known ODE for a pendulum $\theta'' = -gsin\theta$. However, such a simple elimination procedure is usually impossible in more general situations.

Additional examples of real-life DAE systems, including multibody mechanical systems, an electrical circuit, and a prescribed path control problem can be found in Brenan. It should be noted that the constraint in mechanics, e.g. in the pendulum example, is physical, while the constraint in other problems such as a prescribed path problem is not physical but rather part of the performance specifications.

Importance Of DAE's

DAEs are a generalization of an ordinary differential equations (ODEs)

$$x' = f(t, x),$$

for which there is a very rich literature for both mathematical theory and numerical solution. While the standard-form ODE can be written as a DAE, the more general DAE form admits problems that can be quite different from a standard-form ODE. The class of DAEs includes problems exhibiting fundamental mathematical properties that are different from those of ODEs, and also pose additional challenges for their numerical solution. On the other hand, implicit DAE models are formulated in a more natural way than explicit ones, as the above examples demonstrate. It is easier to derive a complex DAE model so it is highly desirable to be able to work with the DAE model if possible.

Index and mathematical structure

Index

Index is a notion used in the theory of DAEs for measuring the distance from a DAE to its related ODE. The index is a nonnegative integer that provides useful information about the mathematical structure and potential complications in the analysis and the numerical solution of the DAE. In general, the higher the index of a DAE, the more difficulties one can expect for its numerical solution. There are different index definitions: Kronecker index (for linear constant coefficient DAEs), differentiation index, perturbation index, tractability index , geometric index, and strangeness index . On simple problems they are identical. On more complicated nonlinear and fully implicit systems they can be different. In fact, the index can become a local concept with different values in different regions. The index may even be undefined at so-called singular points, which typically exhibit impasse phenomena.

Since a DAE involves a mixture of differentiations and integrations, one may hope that differentiating the constraints (in a semi-explicit DAE system) and substituting as needed from the differential equations, repeatedly if necessary, will yield an explicit ODE system for all unknowns. The solutions of the DAE are those solutions of this ODE which are in a subset called the solution manifold. The number of repetitions needed for this transformation is called the differential index of the DAE. Thus, ODEs have index 0 . Consider some simple examples.

Example: Let q(t) be a given, smooth function, and consider the following problems for x(t) .

- The scalar equation

$$x(t) = q(t)$$

is a (trivial) index-1 DAE, since it takes one differentiation to obtain the ODE $x' = q'(t)$.

- For the system

$$x_1 = q(t)$$

$$x_2 = x_1',$$

one differentiates the first equation to get $x_2 = x_1' = q'(t)$ and then $x_2' = x_1'' = q''(t)$. The index is 2 since two differentiations are needed.

Note that whereas m initial or boundary value conditions must be given to specify the solution of a first order ODE of size m, for the simple DAEs in the above example the solution is completely determined by the right hand side and there is only one initial condition that is consistent. General DAE systems usually include also some ODE subsystems. Thus, a DAE system will in general have l *degrees of freedom*, where l is anywhere between 0 and m. However, in general it may be difficult, or at least not immediately obvious, to determine which l pieces of information are needed to determine the solution. Initial or boundary condition which are specified for the DAE must be consistent. In other words, they must satisfy the constraints and possibly even the differentiated constraints of the system. For example, an initial condition on the index-1 system $x(t) = q(t)$ (which is needed if one writes it as an ODE) must satisfy $x_1(0) = q(0)$. For the index-2 system $\begin{matrix} x_1 = q(t) \\ x_2 = x_1', \end{matrix}$, the situation is somewhat more complicated. Not only must any solution satisfy the obvious constraint $x_1(t) = q(t)$, there is also a hidden constraint $x_2(t) = q'(t)$, so the only consistent initial conditions are $x_1(0) = q(0), x_2(0) = q'(0)$. This is an important difference between index-1 and *higher-index* (index greater than 1) DAEs. Higher-index DAEs include some hidden constraints.

Consider again the semi-explicit DAE $\begin{matrix} y' &=& f(t,y,z) \\ 0 &=& g(t,y,z) \end{matrix}$. The index is one if $\partial g / \partial z$ is nonsingular because in that case, one differentiation of the algebraic equation yields z'. For the semi-explicit index-1 DAE one can distinguish between differential variables y whose derivative appears in the equations and algebraic variables z whose derivative does not explicitly appear. It is also worth noting that the algebraic variables may be less smooth than the differential variables by one derivative, e.g. the algebraic variables may be non-differentiable.

In the general case $F(t,x,x') = 0$, $t_0 \leq t \leq t_f$, each component of the solution x may be a mix of differential and algebraic components, which makes the qualitative analysis as well as the numerical solution of such high-index problems much harder and riskier. The semi-explicit form is decoupled in this sense. Any DAE $F(t,x,x') = 0$, $t_0 \leq t \leq t_f$, can be written in the semi-explicit form by introducing a new variable $z = x'$. However, the index of the new DAE is increased by one. Finally, it is important to note, as the following example illustrates, that in general the index may depend also on a particular solution and not only on the form of the DAE.

Example: Consider the DAE system for $x = (x_1, x_2, x_3)^T$

$$x_1' = x_3$$
$$0 = x_2(1 - x_2)$$
$$0 = x_1 x_2 + x_3(1 - x_2) - t$$

The second equation has two solutions $x_2 = 0$ and $x_2 = 1$. If the continuity of x_2 is given, then x_2 does not switch between these two values. It is easy to see that if $x_2 = 0$, then the system is in semi-explicit form and has index-1, while for the case $x_2 = 1$, the system has index-2 and unlike the index-1 case, no initial value of x_1 is required.

Now if one replaces the algebraic equation involving x_2 by $x_2' = 0$, then the index of the new DAE system depends on the initial condition. If $x_2(0) = 1$ the index is 2, otherwise the index is 1.

Special DAE Forms

The general DAE system $F(t,x,x') = 0$, $t_0 \leq t \leq t_f$, can include problems which are not well-defined in mathematical sense, as well as problems which will result in failure for any direct discretization method . Fortunately, many of the higher-index problems encountered in practice can be expressed as a combination of more restrictive structures of ODEs coupled with constraints. One of the more important classes of systems are the *Hessenberg forms* and are given below.

- Hessenberg index-1

$$y' = f(t,y,z)$$

$$0 = g(t,y,z)$$

where the Jacobian g_z is assumed to be nonsingular for all t. This is just a semi-explicit index-1 DAE system mentioned above. Semi-explicit index-1 DAEs are very closely related to implicit ODEs. After solving for z in the algebraic equation (using the implicit function theorem, it can be done in principle), substituting z into the differential equation yields the so-called underlying ODE in y (although no uniqueness is guaranteed). However, for various reasons, this procedure is not always recommended in practice for numerical solution.

- Hessenberg index-2

$$y' = f(t,y,z)$$

$$0 = g(t,y),$$

Where $g_y f_z$ is assumed to be nonsingular for all t. Note that the algebraic variable z is absent from the second equation. This is a pure index-2 DAE and all algebraic variables play the role of index-2 variables. An example arising from modeling incompressible fluid flow by discretized Navier-Stokes equations is given in Ascher.

Numerical Solution

Numerical approaches for the solution of DAEs can be divided into roughly two classes: (i) direct discretizations of the given system and (ii) methods which involve a reformulation (e.g. index reduction), combined with a discretization. The desire for as direct a discretization as possible arises because a reformulation may be costly, it may require more input from the user, and it may involve more user intervention. The reason for the popularity of reformulation approaches is that, as it turns out, direct discretizations are limited in their utility essentially to index-1, index-2 Hessenberg, and index -3 Hessenberg DAE systems.

Fortunately, many DAEs encountered in practical applications are either index-1 or, if higher-index, can be expressed as a simple combination of Hessenberg systems. However, some worse-case difficulties may occur and the most robust direct applications of numerical ODE methods do not always work as one might hope, even for these restricted classes of problems. For a DAE of index greater than two it is usually best to use one of the index-reduction techniques to solve the problem in a lower-index form.

Differential equations such as

$$y' = f(t,y,z)$$

$$\varepsilon z' = g(t,y,z)$$

where ε is a small parameter are called singularly perturbed ODE systems. When the parameter ε is set to be 0 , above equation becomes the DAE $\begin{aligned} y' &= f(t,y,z) \\ 0 &= g(t,y,z) \end{aligned}$. Since the system is above equation (in general) very stiff for small ε , it is natural to consider methods for stiff ODEs for the direct discretization of the limit DAE, and for DAEs of the form $F(t,x,x') = 0, \ t_0 \leq t \leq t_f$, in general. In particular, ODE methods which have stiff decay such as BDF and Radau collocation methods are useful.

Numerical Methods/Direct Discretization

- Backward Euler method/Example of instability

The idea of a direct discretization is simple: approximate x and x' by a discretization formula like multistep methods or Runge-Kutta methods. As an illustration of the use of direct discretization, consider the backward Euler method, the simplest method which has the stiff decay property. Applying the backward difference formula to x' in $F(t,x,x') = 0, \ t_0 \leq t \leq t_f$, a system of N nonlinear equations for x_n

$$F(t_n, x_n, \frac{x_n - x_{n-1}}{h_n}) = 0 \text{ for } n = 1, 2, \dots,$$

results. Here t_n are the time points where we are computing the approximation, xn is the approximation of $x(t_n)$, and $h_n = t_n - t_{n-1}$ is the time step or step size. Once this nonlinear equation system has been recursively solved, a numerical solution for $F(t,x,x') = 0, \ t_0 \leq t \leq t_f$, is obtained. This method works well for index-1 DAEs, and is particularly appropriate for stiff index-1 DAEs, as well as for stiff ODEs.

For higher-index DAEs this simple method, as well as other methods, does not always work. In the worst case, there are simple higher-index DAE systems with well-defined and stable solutions for which the backward Euler method, and in fact all other multistep and Runge-Kutta methods, are unstable or not even applicable. In multibody mechanics simulations the use of stable versus unstable numerical methods is visualized. Some practical difficulties may occur as well during solving the nonlinear system $F(t_n, x_n, \frac{x_n - x_{n-1}}{h_n}) = 0$ for $n = 1, 2, \dots,$ for x_n given x_{n-1} . The solution must be accomplished with a type of iterative numerical method such as a Newton method. These technical difficulties are why, in general, a direct discretization of fully implicit DAEs of index higher than one is not recommended. For fully implicit index-1 and semi-explicit index-2 DAEs, it has been shown that the Backward Euler method is first-order accurate, stable and convergent.

- BDF and general linear multistep methods

Euler is only a first-order method. To get a more accurate solution without taking smaller steps, a higher order method is needed. The constant step-size BDF method applied to a general nonlinear DAE of the form $F(t,x,x')=0, \quad t_0 \le t \le t_f$, is given by

$$F\left(t_n, x_n, \frac{1}{\beta_0 h}\sum_{j=0}^{k}\alpha_j x_{n-j}\right)=0,$$

where β_0 and $\alpha_j, j=0,1,...,k$, are the coefficients of the BDF method. It has been shown that the k-step BDF method of fixed step-size h is convergent of order $0\left(h^k\right)$ if all initial values are correct to $0\left(h^k\right)$, and if the Newton iteration on each step is solved to accuracy $0(h^{k+1})$. For general linear multistep methods, similar convergence results have been established, provided that the coefficients of the multistep methods satisfy a set of order conditions which is in addition to the order conditions for ODEs, to attain order greater than 2. These extra conditions are satisfied by BDF methods.

- Radau collocation and implicit Runge-Kutta methods

The s-stage implicit Runge-Kutta method applied to the general nonlinear DAE of the form $F(t,x,x')=0, \quad t_0 \le t \le t_f$, is given by

$$F(t_{n-1}+c_i h, X_{ni}, K_{ni})=0,$$

$$X_{ni}=x_{n-1}+h\sum_{j=1}^{s}\alpha_{ij}k_j, \quad i=1,2,...,s,$$

And

$$x_n = x_{n-1}+h\sum_{i-1}^{s}b_i k_{ni},$$

Where $c_i, a_{ij}, b_i, i, j=1,2,...,s$, are the coefficients of the Runge-Kutta method. We assume in addition that the matrix $A=(a_{ij})$ is nonsingular.

For the semi-explicit DAE $\begin{aligned} y' &= f(t,y,z) \\ 0 &= g(t,y,z) \end{aligned}$, the formula $X_{ni}=x_{n-1}+h\sum_{j=1}^{s}\alpha_{ij}k_j, \quad i=1,2,...,s$, for the internal stages reads

$$k_{ni}=f\left(t_{n-1}+c_i h, Y_{ni}, Z_{ni}\right),$$

$$Y_{ni}=y_{n-1}+h\sum_{j=1}^{s}a_{ij}k_j,$$

$$g\left(t_{n-1}+c_i h, Y_{ni}, Z_{ni}\right)=0. \quad i=1,2,...,s.$$

It is possible to avoid the quadrature step $x_n = x_{n-1} + h \sum_{i-1}^{s} b_i k_{ni}$, for the algebraic variables z by making the use of stiffly accurate methods, i.e. Runge-Kutta methods satisfying $b_j = a_{sj}, j = 1, 2, ..., s$, . Instead of $x_n = x_{n-1} + h \sum_{i-1}^{s} b_i k_{ni}$, one simply sets $y_n = Y_{ns}$. As was the case for general multistep methods, there are additional order conditions which the method coefficients must satisfy for the method to attain order greater than 2. For Runge-Kutta methods, the requirement of extra order conditions arises even for semi-explicit index-1 DAEs.

It should also be noted that the implementation of direct discretization methods for DAEs faces some additional practical difficulties such as how to obtain a consistent set of initial conditions, a treatment of the ill-conditioning of iteration matrix, and finally, error estimation and stepsize control for index-2 Hessenberg DAEs.

For some special classes of DAEs such as semi-explicit DAEs in Hessenberg form and ODEs on manifolds, in particular for DAEs arising in multibody mechanics, there exist very efficient and robust numerical methods called stabilized or projected methods. The main idea is first to discretize the differential equations by an appropriate numerical ODE method. This step is followed by a post-stabilization or a coordinate projection step to bring the numerical solution closer to satisfying the constraint.

Universal Differential Equation

A universal differential equation (UDE) is a nontrivial differential-algebraic equation with the property that its solutions approximate to arbitrary accuracy any continuous function on any interval of the real line.

Rubel found the first known UDE by showing that, given any continuous function $\phi : \mathbb{R} \to \mathbb{R}$ and any positive continuous function $\in : \mathbb{R} \to \mathbb{R}^+$, there exists a C^∞ solution y of

$$3y'^4 \, y'' \, y'''^2 - 4y'^4 \, y''^2 \, y'''' + 6y'^3 \, y''^2 \, y''' \, y'''' +$$
$$24y'^2 \, y''^4 \, y'''' - 12 \, y'^3 \, y'' \, y'''^3 - 29 \, y'^2 \, y'''^3 \, y''^2 + 12 \, y''^7 = 0$$

such that

$$\left| y(t) - \phi(t) \right| < \in(t)$$

for all $t \in \mathbb{R}$.

Duffin found two additional families of UDEs,

$$n^2 \, y'''' \, y'^2 + 3 \, n(1 - n) \, y''' \, y'' \, y' + \left(2 \, n^2 - 3 \, n + 1\right) y''^3 = 0$$

and

$$n \, y'''' \, y'^2 + (2 - 3n) \, y''' \, y'' \, y' + 2(n - 1) \, y''^3 = 0 \,,$$

Whose solutions are C^n for $n > 3$.

Briggs found a further family of UDEs given by

$$y'''' \, y'^2 - 3 \, y''' \, y'' \, y' + 2\left(1 - n^{-2}\right) y''^3 = 0 \text{ for } n > 3.$$

Stochastic Differential Equation

Stochastic differential equations (SDE's) play an important role in stochastic modeling. For example, in Economics solutions of the SDE's considered below are used to model share prices. In Biology solutions of stochastic partial differential equations (not considered here) describe sizes of populations.

Stochastic differential equations are (for us) a formal abbreviation of integral equations as described now. We fix a stochastic basis $\left(\Omega, \mathcal{F}, \mathbb{P}; \left(\mathcal{F}_t\right)_{t \geq 0}\right)$ which satisfies the usual P assumptions and an $\left(\mathcal{F}_t\right)_{t \geq 0^-}$ Brownian motion $B = \left(B_t\right)_{t \geq 0}$.

- Let $x_0 \in \mathbb{R}$, $D \subseteq \mathbb{R}$ be an open set, and $\sigma, a : [0, \infty) \times D \to \mathbb{R}$ be continuous. A continuous and adapted stochastic process $X = \left(X_t\right)_{t \geq 0}$ is a solution of the stochastic differential equation (SDE)

$$dX_t = \sigma\left(t, X_t\right) dB_t + a\left(t, X_t\right) dt \text{ with } X_0 = x_0$$

provided that the following conditions are satisfied:

(i) $X_t(\omega) \in D$ for all $t \geq 0$ and $\omega \in \Omega$.

(ii) $X_0 \equiv x_0$.

(iii) $X_t = x_0 + \int_0^t \sigma\left(u, X_u\right) dB_u + \int_0^t a\left(u, X_u\right) du$ for $t \geq 0$ a.s.

Let us give some examples of SDE's.

Example: (Brownian motion). A solution of

$$dX_t = dB_t \text{ and } X_0 = 0$$

is the Brownian motion itself $B = \left(B_t\right)_{t \geq 0}$ since $B_t = \int_0^t 1 dB_u$. We can take $D = \mathbb{R}$.

Example: (Geometric Brownian motion with drift). Letting $X_t := x_0 e^{cB_t + bt}$ with $x_0, b, c \in \mathbb{R}$ we obtain by $IT\hat{o}'s$ formula that, a.s.,

$$\begin{aligned}
X_t &= x_0 + \int_0^t c X_u dB_u + \int_0^t b X_u du + \frac{1}{2} \int_0^t c^2 X_u du \\
&= x_0 + \int_0^t c X_u dB_u + \int_0^t \left[b + \frac{1}{2} c^2\right] X_u du \\
&= x_0 + \int_0^t \sigma X_u dB_u + \int_0^t a X_u du
\end{aligned}$$

With

$$\sigma := c,$$

$$a := b + \frac{1}{2}c^2.$$

Going the other way round by starting with a and σ, we get that

$$c = \sigma,$$

$$b = a - \frac{1}{2}\sigma^2.$$

Consequently, the SDE

$$dX_t = \sigma X_t dB_t + a X_t dt \text{ with } X_0 = x_0$$

is solved by

$$X_t = x_0 e^{\sigma B_t + \left(a - \frac{1}{2}\sigma^2\right)t}$$

We may use $D = \mathbb{R}$ for $\sigma(t, x) := \sigma x$ and $a(t, x) := ax$.

The following examples only provide the formal SDE's.

Example: (Ornstein-Uhlenbeck process). Here one considers the SDE

$$dX_t = -cX_t dt + \sigma dB_t \text{ with } X_0 = x_0.$$

Example: (Vasicek interest rate model). Here one considers that

$$dr_t = a(b - r_t)dt + \sigma dB_t \text{ with } r_0 \geq 0,$$

$\sigma \geq 0$, and $a, b > 0$ models an interest rate in Stochastic Finance. The problem with this model is that r_t might be negative if $\sigma > 0$. If $\sigma = 0$, then one gets as one solution

$$r_t = r_0 e^{-at} + b\left(1 - e^{-at}\right)$$

So that the meaning of a and b become more clear: the interest rate moves from its initial value r_0 to the value b as $t \to \infty$ with a speed determined by the parameter a. If $\sigma > 0$ one tries to add a random perturbation to that.

Both, the Ornstein-Uhlenbeck process and the process in the Vasicek interest rate model are Gaussian processes since the diffusion coefficient is not random. Moreover, the Vasicek interest rate model process is a generalization of the Ornstein-Uhlenbeck process.

The drawback of a negative interest rate in the Vasicek model can be removed by the following model:

Example: (Cox-Ingersoll-Ross Model). For $a, b > 0$ and $\sigma \geq 0$ one proposes the SDE

$$dr_t = a(b - r_t)dt + \sigma \sqrt{r_t}dB_t \text{ with } r_0 > 0.$$

The difference to the Vasicek interest rate model is that the factor $\sqrt{r_t}$ is added in the diffusion part. This guarantees that the fluctuation is getting smaller if r_t is close to zero. In fact, the parameters can be adjusted that the trajectories stay positive .

Instead of considering the interest rate r_0 as initial condition one can take into the account the whole interest curve as anticipated by the market at time $t = 0$ as initial condition. This yields to a considerably more complicated model, the Heath-Jarrow-Morton model.

Example: (Heath-Jarrow-Morton model). We assume that $f(s, t)$ stands for the instantaneous interest rate at time t as anticipated by the market at time s with $0 \leq s \leq t < \infty$. In particular, $r_t = f(t, t)$ is the interest rate at time t. Now one considers the equation

$$f(t, u) = f(0, u) + \int_0^t \alpha(v, u)dv + \int_0^t \sigma(f(v, u))dB_v$$

With $f(0, u) = \Phi(u)$.

Strong Uniqueness of SDE's

Lemma (Gronwall): Let $A, B, T \geq 0$ and $f : [0, T] \rightarrow [0,\infty)$ be a continuous function such that

$$f(t) \leq A + B \int_0^t f(s)ds$$

for all $t \in [0, T]$. Then one has that $f(T) \leq Ae^{BT}$.

Proof. Letting $g(t) := e^{-Bt} \int_0^t f(s)ds$ ds we deduce

$$g'(t) = -Be^{-Bt} \int_0^t f(s)ds + e^{-Bt}f(t)$$

$$= e^{-Bt}\left(f(t) - B\int_0^t f(s)ds\right) \leq Ae^{-Bt}$$

and

$$g(T) = \int_0^T g'(t)dt \leq A \int_0^T e^{-Bt}dt = \frac{A}{B}\left(1 - e^{-BT}\right).$$

Consequently,

$$f(T) \leq A + B \int_0^T f(t)dt = A + Be^{BT}g(T)$$

$$\leq A + Be^{BT}\frac{A}{B}\left(1 - e^{-BT}\right) = Ae^{BT}$$

Proposition: (Strong uniqueness). Suppose that for all $n = 1, 2, \ldots$ there is a constant $c_n > 0$ such that

$$\left| \sigma(t, x) - \sigma(t, y) \right| + \left| a(t, x) - a(t, y) \right| \leq c_n \left| x - y \right|$$

for $|x| \leq n$, $|y| \leq n$, and $t \geq 0$. Assume that $(X_t)_{t \geq 0}$ and $(Y_t)_{t \geq 0}$ are solutions of the SDE $dX_t = \sigma(t, X_t) dB_t + a(t, X_t) dt$ with $X_0 = x_0$. Then

$$\mathbb{P}\left(X_t = Y_t , t \geq 0 \right) = 1.$$

Proof. We use the stopping times

$$\sigma_n := \inf \left\{ t \geq 0 : |X_t| \geq n \right\} \text{ and } \tau_n := \inf \left\{ t \geq 0 : |Y_t| \geq n \right\}$$

where we assume that $n > |x_0|$. Letting $\rho_n := \min \left\{ \sigma_n, \tau_n \right\}$ we obtain, a.s., that

$$X_{t \wedge \rho_n} - Y_{t \wedge \rho_n}$$

$$= \int_0^{t \wedge \rho_n} \left[a(u, X_u) - a(u, Y_u) \right] du + \int_0^{t \wedge \rho_n} \left[\sigma(u, X_u) - \sigma(u, Y_u) \right] dB_u .$$

Hence

$$\mathbb{E} \left| X_{t \wedge \rho_n} - Y_{t \wedge \rho_n} \right|^2 \leq 2\mathbb{E} \left| \int_0^{t \wedge \rho_n} \left[a(u, X_u) - a(u, Y_u) \right] du \right|^2$$

$$+ 2\mathbb{E} \left| \int_0^{t \wedge \rho_n} \left[\sigma(u, X_u) - \sigma(u, Y_u) \right] dB_u \right|^2$$

$$\leq 2t\mathbb{E} \int_0^{t \wedge \rho_n} \left| a(u, X_u) - a(u, Y_u) \right|^2 du$$

$$+ 2\mathbb{E} \int_0^{t \wedge \rho_n} \left[\sigma(u, X_u) - \sigma(u, Y_u) \right]^2 du$$

$$\leq (2t + 2)c_n^2 \, \mathbb{E} \int_0^{t \wedge \rho_n} \left| X_u - Y_u \right|^2 du$$

$$\leq (2t + 2)c_n^2 \, \mathbb{E} \int_0^t \left| X_{u \wedge \rho_n} - Y_{u \wedge \rho_n} \right|^2 du.$$

Now fix $T > 0$. The above computation gives

$$\mathbb{E} \left| X_{t \wedge \rho_n} - Y_{t \wedge \rho_n} \right|^2 \leq (2T + 2)c_n^2 \int_0^t \mathbb{E} \left| X_{u \wedge \rho_n} - Y_{u \wedge \rho_n} \right|^2 du$$

for $t \in [0, T]$. For

$$f(t) := \mathbb{E}\left|X_{t \wedge \rho_n} - Y_{t \wedge \rho_n}\right|^2$$

we may apply Gronwall's lemma. The function f is continuous since for $t_k \to t$ one gets

$$
\begin{aligned}
\lim_k f(t_k) &= \lim_k \mathbb{E}\left|X_{t_k \wedge \rho_n} - Y_{t_k \wedge \rho_n}\right|^2 \\
&= \mathbb{E}\lim_k \left|X_{t_k \wedge \rho_n} - Y_{t_k \wedge \rho_n}\right|^2 \\
&= \mathbb{E}\left|X_{t \wedge \rho_n} - Y_{t \wedge \rho_n}\right|^2 \\
&= f(t)
\end{aligned}
$$

by dominated convergence as a consequence of (for example)

$$\mathbb{E}\sup_{t \in [0,T]}\left|X_{t \wedge \rho_n}\right|^2 \le n^2$$

and the continuity of the processes X and Y. Exploiting Gronwall's lemma with $A := 0$ and $B := (2T + 2)c_n^2$ yields

$$f(T) \le Ae^{BT} = 0 \text{ and } \mathbb{E}\left|X_{t \wedge \rho_n} - Y_{t \wedge \rho_n}\right|^2 = 0.$$

Since

$$\lim_n \rho_n = \infty$$

because X and Y are continuous processes, we get by Fatou's lemma that

$$\mathbb{E}\left|X_t - Y_t\right|^2 = \mathbb{E}\liminf_n \left|X_{t \wedge \rho_n} - Y_{t \wedge \rho_n}\right|^2 \le \liminf_n \mathbb{E}\left|X_{t \wedge \rho_n} - Y_{t \wedge \rho_n}\right|^2 = 0.$$

Hence $\mathbb{P}\left(X_t = Y_t\right) = 1$ and, by the continuity of X and Y_t,

$$\mathbb{P}\left(X_t = Y_t, t \ge 0\right) = 1$$

Sometimes the assumptions of the above criteria are too strong. There is a nice extension:

Proposition: (Yamada-Tanaka). Suppose that

$$\sigma, a : [0,\infty) \times \mathbb{R} \to \mathbb{R}$$

are continuous such that

$$\left|\sigma(t, x) - \sigma(t, y)\right| \le h(|x - y|),$$

$$\left|a(t, x) - a(t, y)\right| \le K(|x - y|)$$

for x, $y \in \mathbb{R}$, where $h : [0,\infty) \to [0,\infty)$ is strictly increasing with $h(0) = 0$ and $K : [0,\infty) \to \mathbb{R}$ is strictly increasing and concave with $K(0) = 0$, such that

$$\int_0^\varepsilon \frac{du}{K(u)} = \int_0^\varepsilon \frac{du}{h(u)^2} = \infty$$

for all $\varepsilon > 0$. Then any two solutions of $dX_t = \sigma(t, X_t)dB_t + a(t, X_t)dt$ with $X_0 = x_0$ are indistinguishable.

Example: One can take $h(x) := x^\alpha$ for $\alpha \geq \dfrac{1}{2}$.

For $\alpha = 1/2$ we have in the Cox-Ingersoll-Ross model that $\sigma(t, x) = \sqrt{|x|}$. This implies that

$$\left|\sigma(t, x) - \sigma(t, y)\right| \leq \sigma |\sqrt{x} - \sqrt{y}| \leq \sigma\sqrt{|x - y|}.$$

However, there is also the following example:

Example

Let $\sigma : \mathbb{R} \to [0, \infty)$ be continuous such that

(i) $\sigma(x_0) = 0$,

(ii) $\int_{x_0-\varepsilon}^{x_0+\varepsilon} \dfrac{dx}{\sigma^2(x)} < \infty$ and $\sigma(x) \geq 1$ if $|x - x_0| > \varepsilon$ for some $\varepsilon > 0$.

Then the SDE

$$dX_t = \sigma(X_t)dB_t \text{ with } X_0 = x_0$$

has infinitely many solutions.

Existence of Strong Solutions of SDE's

Proposition: Suppose that σ, $a : [0, \infty) \times \mathbb{R} \to \mathbb{R}$ are continuous such that

$$\left|\sigma(t, x) - \sigma(t, y)\right| + \left|a(t, x) - a(t, y)\right| \leq K|x - y|$$

for all x, $y \in \mathbb{R}$ and some $K > 0$. Then there exists a solution $X = (X_t)_{t\geq 0}$ to the SDE. Moreover $dX_t = \sigma(t, X_t)dB_t + a(t, X_t)dt$ with $X_0 = x_0$, we have

$$\mathbb{E} \sup_{t \in [0,T]} |X_t|^p < \infty$$

for all $T > 0$ and $0 < p < \infty$, and

$$\|X_t\|_p \leq \sqrt{2} \left[|x_0| + 1\right]e^{\frac{K^2}{T} T\left[\alpha_p + \sqrt{T}\right]^2}$$

for $2 \leq p < \infty$, where $\alpha_p > 0$ is the constant from the Burkholder-DavisGundy inequality and

$$K_T := max\left\{K, sup_{t\in[0,T]}\left[\left\|\sigma(t,0)\right\| + |a(t,0)|\right]\right\}.$$

Proof: (a) Let $p \geq 2$ and $T > 0$. By defining

$$\|f\|_{L_p^{C[0,T]}} := \left(\int_\Omega \|f(\omega)\|_{C[0,T]}^p \, d\mathbb{P}(\omega)\right)^{1/p},$$

where $\|f\|_{C[0,T]} := sup_{t\in[0,T]} |f_t|$, we get that

$$L_p^{C[0,T]}(\Omega,\mathbf{F},\mathbb{P}) := \left\{f:\Omega \to C[0,T]:\|f\|_{L_p^{C[0,T]}} < \infty\right\}$$

is a complete normed space, i.e. a Banach space. We define the sequence of processes $X^{(k)} = \left(X_t^{(k)}\right)_{t\geq 0}$ by

$$X_t^{(0)} := x_0,$$

$$X_t^{(k+1)} := x_0 + \int_0^t \sigma\left(u, X_u^{(k)}\right)dB_u + \int_0^t a\left(u, X_u^{(k)}\right)du$$

and observe that $X^{(k)} \in L_p^{C[0,T]}(\Omega,\mathbf{F},\mathbb{P})$ for all $k \in \mathbb{N}$ by induction: For $X_t^{(0)} = x_0$ this is evident, so let us assume $X^{(k)} \in L_p^{C[0,T]}(\Omega,\mathbf{F},\mathbb{P})$. Then, by the triangle-inequality,

$$\left\|X^{(k+1)}\right\|_{L_p^{C[0,T]}} \leq \left\|X^{(k+1)} - X^{(k)}\right\|_{L_p^{C[0,T]}} + \left\|X^{(k)}\right\|_{L_p^{C[0,T]}}$$

and therefore $X^{(k+1)} \in L_p^{C[0,T]}(\Omega,\mathbf{F},\mathbb{P})$. We decompose

$$X_t^{(k+1)} - X_t^{(k)} = \int_0^t \sigma\left(u, X_u^{(k)}\right) - \sigma\left(u, X_u^{(k-1)}\right) dB_u$$

$$+ \int_0^t \left[a\left(u, X_i^{(k)}\right) - a\left(u, X_u^{(k-1)}\right)\right] du$$

$$=: M_t + C_t .$$

Now

$$\mathbb{E} \sup_{t\in[0,T]} |C_t|^p \leq T^{p-1}K^p \int_0^T \mathbb{E}\left|X_u^{(k)} - X_u^{(k-1)}\right|^p du$$

by Hölder's inequality and

$$\mathbb{E} \sup_{t\in[0,T]} |M_t|^p \leq \alpha_p^p \mathbb{E}\left|\int_0^T \left|\sigma\left(u, X_u^{(k)}\right) - \sigma\left(u, X_u^{(k-1)}\right)\right|^2 du\right|^{p/2}$$

$$\leq \alpha_p^p K^p T^{\frac{p-2}{2}} \int_0^T \mathbb{E}\left|X_u^{(k)} - X_u^{(k-1)}\right|^p du$$

by the Burkholder-Davis-Gundy inequality with constant $\alpha_p > 0$ and Hölder's inequality. Consequently,

$$\left(\mathbb{E} \sup_{t \in [0,T]} \left| X_t^{(k+1)} - X_t^{(k)} \right|^p \right)^{1/p} \leq \left(\mathbb{E} \sup_{t \in [0,T]} |M_t|^p \right)^{1/p} + \left(\mathbb{E} \sup_{t \in [0,T]} |C_t|^p \right)^{1/p}$$

$$\leq L_p \left(\int_0^T \mathbb{E} \left| X_u^{(k)} - X_u^{(k-1)} \right|^p du \right)^{1/p}$$

with $L_p := K[T^{\frac{p-1}{p}} + \alpha_p T^{\frac{p-2}{2p}}]$ where we used the triangle-inequality in $L_p^{C[0,T]}(\Omega, \mathcal{F}, \mathbb{P})$. Iterating the last equation yields to

$$\left(\mathbb{E} \sup_{t \in [0,T]} \left| X_t^{(k+1)} - X_t^{(k)} \right|^p \right) \leq \frac{\left(TL_p^p\right)^k}{k!} \sup_{t \in [0,T]} \mathbb{E} \left| X_t^{(1)} - x_o \right|^p =: \frac{\left(TL_p^p\right)^k}{k!} c_{x_0,T,P}^p$$

and shows that $\left(X^{(k)} \right)_{k=1}^\infty$ is a Cauchy-sequence in $L_p^{C[0,T]}(\Omega, \mathcal{F}, \mathbb{P})$: For $0 \leq k < l < \infty$ we have

$$\left\| X^{(l)} - X^{(k)} \right\|_{L_p^{C[0,T]}} \leq \left\| X^{(l)} - X^{(l-1)} \right\|_{L_p^{C[0,T]}} + \cdots + \left\| X^{(k+1)} - X^{(k)} \right\|_{L_p^{C[0,T]}}$$

$$\leq c_{x_0,T,p} \left[\frac{\left(TL_p^p\right)^{l-1/p}}{\left((l-1)!\right)^{1/p}} + \cdots + \frac{\left(TL_p^p\right)^{k/p}}{\left(k!\right)^{1/p}} \right]$$

$$\leq c_{x_0,T,p} \sum_{i=k}^\infty \frac{\left(TL_p^p\right)^{i/p}}{\left(i!\right)^{1/p}} \to_k 0.$$

Therefore the limit X is in $L_p^{C[0,T]}(\Omega, \mathcal{F}, \mathbb{P})$ and

$$\mathbb{E} \sup_{t \in [0,T]} |X_t|^p < \infty$$

for all $T > 0$ and $2 \leq p < \infty$. By $\|.\|_{p_0} \leq \|.\|_{p_1}$ for $0 < p_0 < p_1 < \infty$ this holds for $0 < p < 2$ as well.

(b) By the uniqueness argument for the strong solutions we also get that

$$\mathbb{P}\left(X_t^{(T_1)} = X_t^{(T_2)} \right) = 1$$

for $t \in [0, \min\{T_1, T_2\}]$ when $X^T = \left(X_t^T \right)_{t \in [0,T]}$ is the solution constructed on $[0, T]$. Hence we may find a continuous and adapted process $X = (X_t)_{t \geq 0}$ such that

$$\mathbb{P}\left(X_t = X_t^{(n)} \right) = 1 \text{ for all } t \in [0, n]$$

which turns out to be our solution.

(c) Now we consider the solved equation and $2 \leq p < \infty$. Similarly as before we deduce, for $t \in [0, T]$, that

$$\|X_t\|_p$$

$$\leq |x_0| + \alpha_p \left[\mathbb{E} \left| \int_0^t |\sigma(u, X_u)|^2 \, du \right|^{P/2} \right]^{\frac{1}{p}} + \left(\mathbb{E} \left| \int_0^t |a(u, X_u)| \, du \right|^p \right)^{\frac{1}{p}}$$

$$\leq |x_0| + \alpha_p K_T \left[\mathbb{E} \left| \int_0^t [1 + |X_u|]^2 \, du \right|^{P/2} \right]^{\frac{1}{p}} +$$

$$K_T \left(\mathbb{E} \left| \int_0^t [1 + |X_u|] \, du \right|^p \right)^{\frac{1}{p}}$$

$$\leq |x_0| + \alpha_p K_T \left| \int_0^t \|1 + |X_u|\|_p^2 \, du \right|^{1/2} + K_T \int_0^t \|1 + |X_u|\|_p \, du$$

$$\leq |x_0| \, K_T [\alpha_p + \sqrt{T}] \left| \int_0^t \|1 + |X_u|\|_p^2 \, du \right|^{1/2}.$$

Consequently,

$$\|1 + |X_t|\|_p^2 \leq 2[|x_0| + 1]^2 + 2K_T^2 [\alpha_p + \sqrt{T}]^2 \int_0^t \|1 + |X_u|\|_p^2 \, du$$

for $t \in [0, T]$. Gronwall's lemma gives

$$\|1 + |X_t|\|_p^2 \leq 2[|x_0| + 1]^2 \, e^{2K_T^2 [\alpha_p + \sqrt{T}]^2}.$$

Remark:

1. From the above proof it follows that we obtain a Gaussian process in case of $\sigma(t, x) = \sigma(t)$ and $a(t, x) = a_1(t)x + a_2(t)$ *with* σ, a_1, a_2 continuous and bounded as the approximating processes $X^{(k)}$ are Gaussian and the L_2-limit of Gaussian random variables is Gaussian as well.

2. From the assumptions of above proposition we get that

$$|\sigma(t, x)| + |a(t, x)| \leq \sup_{t \in [0,T]} [|\sigma(t, 0)| + |a(t, 0)|] + K|x|$$

which is a standard growth condition that is satisfied in our context automatically.

References

- Michiels, Wim; Niculescu, Silviu-Iulian (2007). Stability and stabilization of time-delay systems. An eigenvalue based approach. doi:10.1137/1.9780898718645. ISBN 978-0-89871-632-0

- Differential-equation, science: britannica.com, Retrieved 16 May 2018

- Zeilberger, Doron. A holonomic systems approach to special functions identities. Journal of computational and applied mathematics. 32.3 (1990): 321-368

- Order-and-degree-of-a-differential-equation: toppr.com, Retrieved 10 July 2018

- Partial-differential-equation, science: britannica.com, Retrieved 30 March 2018

- What-is-a-linear-differential-equation: quora.com, Retrieved 14 April 2018

- Briat, Corentin (2015). Linear Parameter-Varying and Time-Delay Systems. Analysis, Observation, Filtering & Control. Springer Verlag Heidelberg. ISBN 978-3-662-44049-0.

- What-is-a-nonlinear-equation: online-courses.vissim.us, Retrieved 31 April 2018

- Differential-algebraic-equations: scholarpedia.org, Retrieved 31 March 2018

- Universal-Differential-Equation: mathworld.wolfram.com, Retrieved 29 April 2018

Differential Equations: First Order and First Degree

The order of a differential equation is determined by the order of the highest derivative, such as, in a first order differential equation the highest term has a first derivative. The power of the highest derivative of a differential equation is called its degree. The aim of this chapter is to explore the fundamentals of first order and first degree differential equations through an analysis of the crucial topics of Separation of variables Bernoulli's equation, homogeneous differential equations, exact differential equations, etc.

Any differential equation of the first order and first degree can be written in the form,

$$\frac{dy}{dx} = F(x, y) \quad or \quad M(x, y)dx + N(x, y)dy = 0$$

Example: The differential equation

$$\frac{dy}{dx} = \frac{x - 3y}{2y - x}$$

can also be written as,

$$(x - 3y)dx + (x - 2y)dy = 0$$

Existence of a solution: The general solution of the equation $dy/dx = g(x, y)$, if it exists, has the form $f(x, y, C) = 0$, where C is an arbitrary constant. Under what circumstances does a general solution exist? We have the following theorem.

Theorem: A general solution of $dy/dx = g(x, y)$ exists over some specified region R of points (x, y) if the following conditions are met:

 a) $g(x, y)$ is continuous and single-valued over R

 b) $\partial g / \partial y$ exists and is continuous at all points of R

The general solution $f(x, y, C) = 0$ of a differential equation $dy/dx = g(x, y)$ over some region R consists of a family of curves, called the integral curves of the differential equation, (one curve for each possible value of C, each curve representing a particular solution), such that through each point in R there passes one and only one curve of the family $f(x, y, C) = 0$.

Differential Equation

$$\frac{dy}{dx} = g(x, y)$$

Associates with each point (x_o, y_o) in the region R a direction

$$m = \frac{dy}{dx}\bigg|_{(x_0 y_0)} = g(x_0, y_0)$$

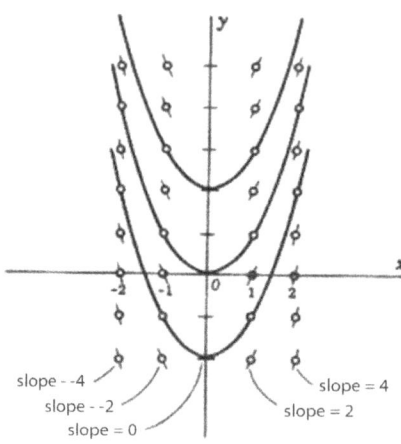

Direction field of dy/dx = 2x

The direction at each point of R is that of the tangent to that curve of the family $f(x, y, C) = 0$ that passes through the point.

A region R in which a direction is associated with each point is called a direction field. In Above Figure, it is shown the direction field and integral curves for the differential equation $dy/dx = 2x$. The general solution of this equation is $y = x^2 + C$. The integral curves are parabolas.

A first order and first degree differential equation can be represented as

$$F(x, y, dy/dx) = 0.$$

Whose solution can be represented as

$$f(x, y) = c$$

If the differential equation represented by $F(x, y, dy/dx) = 0.$ can be divided into two different differential equations and is represented as

$$F_1(x, y, dy/dx) + F_2(x, y, dy/dx) = 0$$

Then the solution of the differential equation represented by $F(x, y, dy/dx) = 0$ can be found using the procedure given below:

Step1. $F_1(x, y, dy/dx) = 0$

And $F_2(x, y, dy/dx) = 0$

Step 2. If $f_1(x, y) = c_1$

And $f_2(x, y) = c_2$

be the solutions of the differential equations represented by $F_1(x, y, dy/dx) = 0$ and $F_2(x, y, dy/dx) = 0$ respectively.

Step 3. Solution of the differential equation represented by $F(x, y, dy/dx) = 0$. is

$$f(x, y) = f_1(x, y) + f_2(x, y) = c_1 + c_2 = c$$

This method can be generalized also. If the differential equation represented by $F(x, y, dy/dx) = 0$. can be divided into n differential equations and is represented as

$$F_1(x, y, dy/dx) + F_2(x, y, dy/dx) + ...+ F_n(x, y, dy/dx) = 0$$

Then the solution of the differential equation represented by $F(x, y, dy/dx) = 0$. is

$$f(x, y) = f_1(x, y) + f_2(x, y) + ...+ f_n(x, y) = c_1 + c_2 + ...+ c_n = c.$$

Example

$$\left(2x y^4 e^y + 2x y^3 + y\right) dx + \left(x^2 y^4 e^y - x^2 y^2 - 3x\right) dy = 0$$

Or $\left(2x e^y + 2x/y + 1/y^3\right) + \left(x^2 e^y - x^2/y^2 - 3x/y^4\right) dy/dx = 0$

Or $\left(e^y - 1/y^2 - 3/(xy^4)\right) dy/dx + (2 e^y/x + 2/(xy) + 1/(x^2 y^3)) = 0$

Or $\left\{\left(e^y - 1/y^2\right) dy/dx + 2/x\left(e^y + 1/y\right)\right\} + \left\{-3/(xy^4) dy/dx + 1/(x^2 y^3)\right\} = 0$

Here differential equation represented by

$$\left(2x y^4 e^y + 2x y^3 + y\right) dx + \left(x^2 y^4 e^y - x^2 y^2 - 3x\right) dy = 0$$

is expressed as sum of two different differential equations i.e.

$$\left(e^y - 1/y^2\right) dy/dx + 2/x\left(e^y + 1/y\right) = 0$$

And $-3/\left(xy^4\right) dy/dx + 1/\left(x^2 y^3\right) = 0$

From $\left(e^y - 1/y^2\right) dy/dx + 2/x\left(e^y + 1/y\right) = 0$,

Let $e^y + 1/y = z$

Or $\left(e^y - 1/y^2\right) dy/dx = dz/dx$ (differentiating w. r. t. x)

Or $dz/dx + (2/x)z = 0$

Or $dz/z + 2(dx/x) = 0$

Or $\ln z + 2\ln x = \ln c1$ (on integration)

Or $zx^2 = c_1$

Or $\qquad (e^y + 1/y) \qquad\qquad x^2 = c_1$

Which is solution of differential equation represented by

$$(e^y - 1/y^2)\, dy/dx + 2/x(e^y + 1/y) = 0.$$

From $-3/(xy^4)\, dy/dx + 1/(x^2 y^3) = 0$,

$-3\, dy/dx + y/x = 0$

Or $-3\, dy/y + dx/x = 0$

Or $-3\ln y + \ln x = \ln c_2$ (on integration)

Or $x/(y^3) = c_2$

Hence, the solution of the differential equation represented by

$$(2x\, y^4\, e^y + 2x\, y^3 + y)\, dx + (x^2\, y^4\, e^y - x^2\, y^2 - 3x)\, dy = 0$$

in the example is

$$(e^y + 1/y)x^2 + x/(y^3) = c$$

Precaution

While solving the differential equations derived from the original equation, Right Hand Side (R.H.S.) of the equation must always be equal to zero and all variable terms must be kept to the Left Hand Side (L.H.S.) of the equation. Transferring the variable terms to the R.H.S. may lead to wrong answer. For example if $3/(xy^4)\, dy/dx + 1/(x^2 y^3) = 0$ of the above mentioned example is expressed as

$3/(xy^4)dy/dx = 1/x^2 y^3$

Or $3\, dy/dx = y/x$

Or $3\, dy/y = dx/x$

Or $3\ln y - \ln x = \ln c_2$ (on integration)

Or $y^3/x = c_2$ which is wrong

Variable Separable, Homogeneous Equation Method, Linear equation of first order and Exact differential equation method are standard methods for solution of first order and first degree differential equations. But some differential equations of first order and first degree are so complex and lengthy that none of the four methods mentioned above can be used to solve the equations easily.

Separation of Variables

Separation of variables is a method of solving ordinary and partial differential equations.

For separation of variables we require that the equation can be (re)written in the form

$$\frac{dy}{dx} = f(x) g(y)$$

Where $f(x)$ is only a function of x and $g(y)$ is only a function of y.

Then the general solution of the first order ODE

$$\frac{dy}{dx} = f(x) g(y)$$

is given by

$$\int \frac{1}{g(y)} dy = \int f(x) dx.$$

Often we need to rewrite the equation so that it is in the correct form. For example, the ODE

$$y \frac{dy}{dx} - 3 = x$$

can be rewritten $y \frac{dy}{dx} = x + 3$ and then as

$$\frac{dy}{dx} = \frac{x + 3}{y} = f(x) g(y)$$

where $f(x) = x + 3$ and $g(y) = \frac{1}{y}$.

Therefore the equation $y \frac{dy}{dx} - 3 = x$ can be solved using separation of variables.

The general approach to solving a first order ODE using separation of variables is as follows:

a) Rewrite (if necessary) the equation in the required form:

$$\frac{dy}{dx} = f(x) g(y)$$

b) Find the general solution for

$$\int \frac{1}{g(y)}\, dy = \int f(x)\, dx$$

c) If boundary conditions are given, solve to find the unique solution.

Example: Solve the ODE

$$\frac{1}{y^2}\frac{dy}{dx} + x^2 = 0$$

subject to the initial condition $y(0) = 2$.

Solution: $\dfrac{dy}{dx} = -x^2 y^2$.

Thus $\displaystyle\int y^{-2}\, dy = \int -x^2\, dx$ and so $-y^{-1} = -\dfrac{1}{3}x^3 + c$

Use the boundary condition $y(0) = 2$, to obtain $c = -\dfrac{1}{2}$ and hence

$$y^{-1} = \frac{1}{3}x^3 + \frac{1}{2}.$$

It is quite often the case (as is true here) that one has an implicit function of y rather than an explicit one. We could rewrite the solution as $y = \dfrac{1}{\dfrac{1}{3}x^3 + \dfrac{1}{2}}$

Example:

Population growth is often modeled by the differential equation

$$\frac{dP}{dt} = kP\left(1 - \frac{P}{K}\right)$$

where P is the population with respect to time t, k is the rate of growth, and K is the carrying capacity of the environment.

Separation of variables may be used to solve this differential equation.

$$\frac{dP}{dt} = kP\left(1 - \frac{P}{K}\right)$$

$$\int \frac{dP}{P\left(1 - \dfrac{P}{K}\right)} = \int k\, dt$$

To evaluate the integral on the left side, we simplify the fraction

$$\frac{1}{P\left(1 - \dfrac{P}{K}\right)} = \frac{K}{P(K - P)}$$

and then, we decompose the fraction into partial fractions

$$\frac{K}{P(K-P)} = \frac{1}{P} + \frac{1}{K-P}$$

Thus we have

$$\int \left(\frac{1}{P} + \frac{1}{K-P} \right) dP = \int k\,dt$$

$$\ln|P| - \ln|K-P| = kt + C$$

$$\ln|K-P| - \ln|P| = -kt - C$$

$$\ln\left|\frac{K-P}{P}\right| = -kt - C$$

$$\left|\frac{K-P}{P}\right| = e^{-kt-C}$$

$$\left|\frac{K-P}{P}\right| = e^{-C}e^{-kt}$$

$$\frac{K-P}{P} = \pm e^{-C}e^{-kt}$$

Let $A = \pm e^{-C}$

$$\frac{K-P}{P} = Ae^{-kt}$$

$$\frac{K}{P} - 1 = Ae^{-kt}$$

$$\frac{K}{P} = 1 + Ae^{-kt}$$

$$\frac{P}{K} = \frac{1}{1 + Ae^{-kt}}$$

$$P = \frac{K}{1 + Ae^{-kt}}$$

Therefore, the solution to the logistic equation is

$$P(t) = \frac{K}{1 + Ae^{-kt}}$$

To find A, let $t = 0$ and $P(0) = P_0$. Then we have

$$P_0 = \frac{K}{1 + Ae^0}$$

Noting that $e^0 = 1$, and solving for A we get

$$A = \frac{K - P_0}{P_0}.$$

Partial Differential Equations

The method of separation of variables is also used to solve a wide range of linear partial differential equations with boundary and initial conditions, such as the heat equation, wave equation, Laplace equation, Helmholtz equation and biharmonic equation.

Example: Homogeneous case

Consider the one-dimensional heat equation. The equation is

$$\frac{\partial u}{\partial t} - \alpha \frac{\partial^2 u}{\partial x^2} = 0$$

The variable u denotes temperature. The boundary condition is homogeneous, that is

$$u\big|_{x=0} = u\big|_{x=L} = 0$$

Let us attempt to find a solution which is not identically zero satisfying the boundary conditions but with the following property: u is a product in which the dependence of u on x, t is separated, that is:

$$u(x,t) = X(x)T(t)$$

Substituting u back into equation $\dfrac{\partial u}{\partial t} - \alpha \dfrac{\partial^2 u}{\partial x^2} = 0$ and using the product rule,

$$\frac{T'(t)}{\alpha T(t)} = \frac{X''(x)}{X(x)}.$$

Since the right hand side depends only on x and the left hand side only on t, both sides are equal to some constant value $-\lambda$. Thus:

$$T'(t) = -\lambda \alpha T(t),$$

and

$$X''(x) = -\lambda X(x).$$

$-\lambda$ here is the eigenvalue for both differential operators, and $T(t)$ and $X(x)$ are corresponding eigenfunctions.

We will now show that solutions for $X(x)$ for values of $\lambda \leq 0$ cannot occur:

Suppose that $\lambda < 0$. Then there exist real numbers B, C such that

$$X(x) = Be^{\sqrt{-\lambda}x} + Ce^{-\sqrt{-\lambda}x}.$$

From $u\big|_{x=0} = u\big|_{x=L} = 0$ we get

$$X(0) = 0 = X(L),$$

and therefore $B = 0 = C$ which implies u is identically 0.

Suppose that $\lambda = 0$. Then there exist real numbers B, C such that

$$X(x) = Bx + C.$$

From $X(0) = 0 = X(L)$, we conclude in the same manner as in 1 that u is identically 0.

Therefore, it must be the case that $\lambda > 0$. Then there exist real numbers A, B, C such that

$$T(t) = Ae^{-\lambda \alpha t},$$

and

$$X(x) = B\sin(\sqrt{\lambda}\,x) + C\cos(\sqrt{\lambda}\,x).$$

From we $X(0) = 0 = X(L)$, get $C = 0$ and that for some positive integer n,

$$\sqrt{\lambda} = n\frac{\pi}{L}.$$

This solves the heat equation in the special case that the dependence of u has the special form of $u(x,t) = X(x)T(t)$.

In general, the sum of solutions to $\dfrac{\partial u}{\partial t} - \alpha \dfrac{\partial^2 u}{\partial x^2} = 0$ which satisfy the boundary conditions $u\big|_{x=0} = u\big|_{x=L} = 0$ also satisfies $\dfrac{\partial u}{\partial t} - \alpha \dfrac{\partial^2 u}{\partial x^2} = 0$ and $u(x,t) = X(x)T(t)$. Hence a complete solution can be given as

$$u(x,t) = \sum_{n=1}^{\infty} D_n \sin\frac{n\pi x}{L}\exp\left(-\frac{n^2\pi^2\alpha t}{L^2}\right),$$

where D_n are coefficients determined by initial condition.

Given the initial condition

$$u\big|_{t=0} = f(x),$$

we can get

$$f(x) = \sum_{n=1}^{\infty} D_n \sin\frac{n\pi x}{L}.$$

This is the sine series expansion of *f(x)*. Multiplying both sides with $\sin\dfrac{n\pi x}{L}$ and integrating over *[0, L]* result in

$$D_n = \frac{2}{L}\int_0^L f(x)\sin\frac{n\pi x}{L}dx.$$

This method requires that the Eigen functions of *x*, here $\left\{\sin\dfrac{n\pi x}{L}\right\}_{n=1}^{\infty}$, are orthogonal and complete. In general this is guaranteed by Sturm-Liouville theory.

Example: Nonhomogeneous case

Suppose the equation is nonhomogeneous,

$$\frac{\partial u}{\partial t} - \alpha \frac{\partial^2 u}{\partial x^2} = h(x,t)$$

with the boundary condition the same as $u\big|_{x=0} = u\big|_{x=L} = 0$.

Expand $h(x,t), u(x,t)$ and $f(x)$ into

$$h(x,t) = \sum_{n=1}^{\infty} h_n(t) \sin \frac{n\pi x}{L},$$

$$u(x,t) = \sum_{n=1}^{\infty} u_n(t) \sin \frac{n\pi x}{L},$$

$$f(x) = \sum_{n=1}^{\infty} b_n \sin \frac{n\pi x}{L},$$

where $h_n(t)$ and b_n can be calculated by integration, while $u_n(t)$ is to be determined.

Substitute $h(x,t) = \sum_{n=1}^{\infty} h_n(t) \sin \frac{n\pi x}{L}$, and $u(x,t) = \sum_{n=1}^{\infty} u_n(t) \sin \frac{n\pi x}{L}$, back to $\frac{\partial u}{\partial t} - \alpha \frac{\partial^2 u}{\partial x^2} = h(x,t)$ and considering the orthogonally of sine functions we get

$$u_n'(t) + \alpha \frac{n^2 \pi^2}{L^2} u_n(t) = h_n(t),$$

which are a sequence of linear differential equations that can be readily solved with, for instance, Laplace transform, or Integrating factor. Finally, we can get

$$u_n(t) = e^{-\alpha \frac{n^2 \pi^2}{L^2} t} \left(b_n + \int_0^t h_n(s) e^{\alpha \frac{n^2 \pi^2}{L^2} s} ds \right).$$

If the boundary condition is nonhomogeneous, then the expansion of $h(x,t) = \sum_{n=1}^{\infty} h_n(t) \sin \frac{n\pi x}{L}$, and $u(x,t) = \sum_{n=1}^{\infty} u_n(t) \sin \frac{n\pi x}{L}$, is no longer valid. One has to find a function v that satisfies the boundary condition only, and subtract it from u. The function $u-v$ then satisfies homogeneous boundary condition, and can be solved with the above method.

Example: Mixed derivatives

For some equations involving mixed derivatives, the equation does not separate as easily as the heat equation did in the first example above, but nonetheless separation of variables may still be applied. Consider the two-dimensional biharmonic equation

$$\frac{\partial^4 u}{\partial x^4} + 2\frac{\partial^4 u}{\partial x^2 \partial y^2} + \frac{\partial^4 u}{\partial y^4} = 0.$$

Proceeding in the usual manner, we look for solutions of the form

$$u(x, y) = X(x)Y(y)$$

and we obtain the equation

$$\frac{X^{(4)}(x)}{X(x)} + 2\frac{X''(x)}{X(x)}\frac{Y''(y)}{Y(y)} + \frac{Y^{(4)}(y)}{Y(y)} = 0.$$

Writing this equation in the form

$$E(x) + F(x)G(y) + H(y) = 0,$$

we see that the derivative with respect to x and y eliminates the first and last terms, so that

$$F'(x)G'(y) = 0,$$

i.e. either $F(x)$ or $G(y)$ must be a constant, say $-\lambda$. This further implies that either $-E(x) = F(x)G(y) + H(y)$ or $-H(y) = E(x) + F(x)G(y)$ are constant. Returning to the equation for X and Y, we have two cases

$$X''(x) = -\lambda_1 X(x)$$
$$X^{(4)}(x) = \mu_1 X(x)$$
$$Y^{(4)}(y) - 2\lambda_1 Y''(y) = -\mu_1 Y(y)$$

and

$$Y''(y) = -\lambda_2 Y(y)$$
$$Y^{(4)}(y) = \mu_2 Y(y)$$
$$X^{(4)}(x) - 2\lambda_2 X''(x) = -\mu_2 X(x)$$

which can each be solved by considering the separate cases for $\lambda_i < 0$, $\lambda_i = 0$, $\lambda_i > 0$ and noting that $\mu_i = \lambda_i^2$.

Curvilinear Coordinates

In orthogonal curvilinear coordinates, separation of variables can still be used, but in some details different from that in Cartesian coordinates. For instance, regularity or periodic condition may determine the eigenvalues in place of boundary conditions.

Matrices

The matrix form of the separation of variables is the Kronecker sum.

As an example we consider the 2D discrete Laplacian on a regular grid:

$$L = \mathbf{D_{xx}} \oplus \mathbf{D_{yy}} = \mathbf{D_{xx}} \otimes \mathbf{I} + \mathbf{I} \otimes \mathbf{D_{yy}},$$

Where, $\mathbf{D_{xx}}$ and $\mathbf{D_{yy}}$ are 1D discrete Laplacians in the x- and y-directions, correspondingly, and $\mathbf{I}$ are the identities of appropriate sizes.

Reducible Equation

Certain first-order differential equation are not separable but can be made separable by a simple change of variables (dependent variable)

The equation of the form $y' = g\left(\dfrac{y}{x}\right)$ can be made separable; and the form is called the R-1 formula.

Step 1. Set $\dfrac{y}{x} = u$, then $y = ux$ (change of variables).

Step 2. Differential $y' = u + xu'$ (product differentiation formula).

Step 3. The original DE $y' = g\left(\dfrac{y}{x}\right) \Rightarrow u + xu' = g(u)$

$$\Rightarrow xu' = g(u) - u \Rightarrow \frac{du}{dx} = \frac{g(u) - u}{x} \Rightarrow \frac{du}{g(u) - u} = \frac{dx}{x}.$$

Step 4. Integrate both sides of the equation.

Step 5. Replace u by y / x.

Example: Solve $2xyy' = y^2 - x^2$

Dividing by x^2, we have

$$2\frac{y}{x}y' - \left(\frac{y}{x}\right)^2 + 1 = 0 \Rightarrow 2u(u + u'x) - u^2 + 1 = 0 \quad \text{by setting } u = \frac{y}{x}$$

$$\Rightarrow 2u^2 + 2uxu' - u^2 + 1 = 0 \Rightarrow 2uxu' + u^2 + 1 = 0$$

$$\Rightarrow -2uxu' = u^2 + 1 \Rightarrow \frac{2udu}{u^2 + 1} = \frac{-dx}{x} \qquad \int \frac{f'(x)}{f(x)} dx = 1n|f(x)| + c$$

$$\Rightarrow \text{In}(1 + u^2) = -\text{In}|x| + c^* \Rightarrow 1 + u^2 = \frac{c}{x}$$

$$\Rightarrow 1 + \left(\frac{y}{x}\right)^2 = \frac{c}{x} \Rightarrow x^2 + y^2 = cx \Rightarrow \left(x - \frac{c}{2}\right)^2 + y^2 = \frac{c^2}{4}.$$

Example: Solve initial value problem

$$y' = \frac{y}{x} + \frac{2x^3 \cos(x^2)}{y}, y\left(\sqrt{\pi}\right) = 0.$$

$$\text{change of variable } u = \frac{y}{x}$$

$$\Rightarrow xu' + u = u + \frac{2x^2 \cos\left(x^2\right)}{u} \Rightarrow uu' = 2x \cos\left(x^2\right)$$

$$\Rightarrow \frac{u^2}{2} = \sin\left(x^2\right) + c \Rightarrow y = ux = x\sqrt{2\sin\left(x^2\right) + 2c}.$$

Since $y\left(\sqrt{x}\right) = 0 \Rightarrow c = 0 \Rightarrow y = x\sqrt{2\sin\left(x^2\right)}.$

Example: Solve $\left(2x - 4y + 5\right)y' + x - 2y + 3 = 0.$

If we set $u = y/x$, then the equation will become no-separable. One way by setting $x - 2y = v.$
Then $y' = \frac{1}{2}(1 - v')$

$$\Rightarrow \left(2v + 5\right)\frac{1}{2}(1 - v') + v + 3 = 0 \quad \Rightarrow \quad v + \frac{5}{2} - vv' - \frac{5}{2}v' + v + 3 = 0$$

$$\Rightarrow 2v + 5 - 2vv' - 5v' + 2v + 6 = 0 \Rightarrow \left(2v + 5\right)v' = 4v + 11$$

$$\Rightarrow \frac{2v + 5}{4v + 11}dv = dx \Rightarrow \frac{1}{2}(1 - \frac{1}{4v + 11}) dv = dx$$

$$\Rightarrow (1 - \frac{1}{4v + 11})dv = 2dx \Rightarrow v - \frac{1}{4} \ln|4v + 11| = 2x + c^*.$$

Since $v = x - 2y$

$$\int \frac{f'(x)}{f(x)} dx = \ln|f(x)| + c$$

$$\Rightarrow x - 2y - \frac{1}{4} \ln|4x - 8y + 11| = 2x + c^*$$

$$\Rightarrow 4x - 8y - \ln|4x - 8y + 11| = 8x - c$$

$$\Rightarrow 4x + 8y + \ln|4x - 8y + 11| = c.$$

Homogeneous Differential Equations

A linear ordinary differential equation of order n is said to be homogeneous if it is of the form

$$a_n\left(x\right)y^{(n)} + a_{n-1}\left(x\right)y^{(n-1)} + \ldots + a_1\left(x\right)y' + a_0\left(x\right)y = 0,$$

where $y' = dy/dx$, i.e., if all the terms are proportional to a derivative of y (or y itself) and there is no term that contains a function of x alone.

However, there is also another entirely different meaning for a first-order ordinary differential equation. Such an equation is said to be homogeneous if it can be written in the form

$$\frac{dy}{dx} + F\left(\frac{y}{x}\right).$$

Such equations can be solved in closed form by the change of variables $u = y/x$ which transforms the equation into the separable equation

$$\frac{dx}{x} = \frac{du}{F(u) - u}.$$

Steps Involved to Solve Homogeneous Differential Equations

There is a method to solve homogeneous first order ODE's (which are not separable)

Step 1: Check your ODE is homogeneous and not separable.

Step 2: Make the substitutions:

$$y = ux \qquad \text{and} \qquad \frac{dy}{dx} = u + x\frac{du}{dx}$$

The second of these results is found by differentiating the first result using the product rule. This eliminates y from your ODE. Making this substitution in a homogeneous ODE always makes a new ODE with variables u and x which is separable.

Step 3: Solve your new ODE using separation of variables.

Step 4: Replace u using $u = \dfrac{y}{x}$.

Step 5: Check your solution by differentiation.

Example:　Solve $\dfrac{dy}{dx} - \dfrac{y}{x} = e^{\frac{y}{x}}$.

Step 1: Replacing y with ty and x with tx in the ODE you get:

$$\frac{dy}{dx} - \frac{ty}{tx} = e^{\frac{ty}{tx}}$$

You can cancel down all the t's and so the ODE is homogeneous. Furthermore you cannot separate the variables here.

Step 2: Next make the substitutions $y = ux$ and $\dfrac{dy}{dx} = u + x\dfrac{du}{dx}$. So:

$$\frac{dy}{dx} - \frac{y}{x} = e^{\frac{y}{x}} \qquad\qquad \text{becomes} \qquad\qquad u + x\frac{du}{dx} - u = e^u$$

Here u cancels on the left-hand side to give: $x\dfrac{du}{dx} = e^u$

Step 3: You can now separate the variables to give:

$$e^{-u}du = \frac{1}{x}dx \qquad \text{integration both sides gives} \qquad -e^{-u} = \text{In } x + c$$

Where the constant c results from collecting together the two integration constants from the indefinite integration on both sides of the equation.

Step 4: Replacing u using $u = \dfrac{y}{x}$ gives.

$$-e^{-\frac{y}{x}} = \ln x + x \qquad \text{general solution of } \dfrac{dy}{dx} - \dfrac{y}{x} = e^{\frac{y}{x}}$$

Step 5: You can check your general solution by using differentiation. Here your general solution is best differentiated by using implicit differentiation which gives:

$$-e^{\frac{y}{x}}\left(\dfrac{1}{x}\dfrac{dy}{dx} - \dfrac{y}{x^2}\right) = \dfrac{1}{x}$$

This can be rearranged to give the original ODE $\dfrac{dy}{dx} - \dfrac{y}{x} = e^{\frac{y}{x}}$

Example: Solve $\dfrac{dy}{dx} = \dfrac{3x^2 + y^2}{xy}$ with the boundary condition $y(1)=1$

Step 1: The right-hand side can be rewritten as:

$$\dfrac{3x}{y} + \dfrac{y}{x}$$

Identifying either x/y or y/x in a homogenous ODE can help you with the next step which involves substitution.

Step 2: Next make the substitutions $y = ux$ and $\dfrac{dy}{dx} = u + x\dfrac{du}{dx}$. So:

$$\dfrac{dy}{dx} = \dfrac{3x}{y} + \dfrac{y}{x} \qquad \text{become} \qquad u + x\dfrac{du}{dx} = \dfrac{3}{u} + u$$

You can subtract u from each side to give: $x\dfrac{du}{dx} = \dfrac{3}{u}$

Step 3: You can now separate the variables to give:

$$udu = -dx \qquad \text{integration both sides gives} \, -\!\!-\!\! = 3\ln x + c$$

Where the constant c results from collecting together the two integration constants from the indefinite integration on both sides of the equation.

Step 4: Replacing u using $u = \dfrac{y}{x}$ gives.

$$\dfrac{y^2}{2x^2} = 3\ln x + c \qquad \text{general solution of } \dfrac{dy}{dx} = \dfrac{3x^2 + y^2}{xy}$$

Step 5: The general solution of this equation is best differentiated using implicit differentiation which gives:

$\dfrac{y}{x^2}\dfrac{dy}{dx} - \dfrac{y^2}{x^3} = \dfrac{3}{x}$ which can be rearranged to give the original ODE.

Remember that y is a function of x, you could write $y(x)$. So $y(1)=1$ tells you that $y=0$ when $x=2$. This is the boundary condition. Using this in the general solution tells you that:

$$\dfrac{y^2}{2x^2} = 3\ln x + c \qquad \text{become} \qquad \dfrac{1}{2} = 0 + c$$

Or, in other words, $c=1/2$. You can now substitute this value back into your general solution to give a solution particular to the given boundary conditions:

$$\dfrac{y^2}{2x^2} = 3\ln x + \dfrac{1}{2} \qquad \text{particular solution of } \dfrac{dy}{dx} = \dfrac{3x^2 + y^2}{xy} \text{ when } y(1)=1$$

A solution like this one, where you have no unknowns, is called a particular solution.

Flow Chart for Solving Homogeneous First Order ODE's

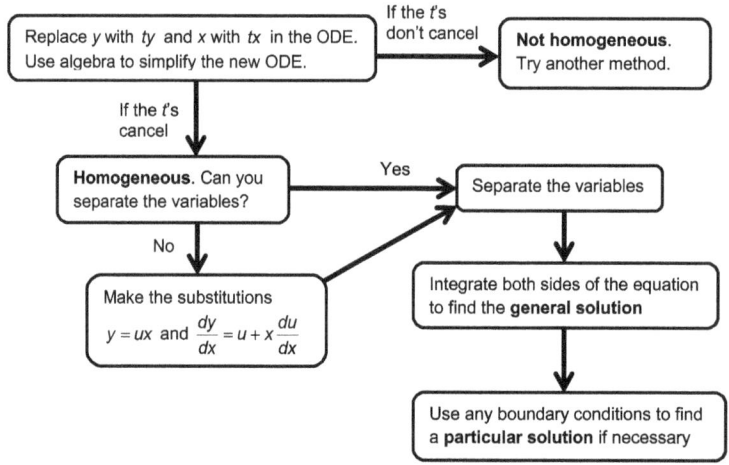

Exercise:

$$(x - y)dx + xdy = 0.$$

Solution:

The coefficients of the differential equations are homogeneous, since for any $a \neq 0$

$$\dfrac{ax - ay}{ax} = \dfrac{x - y}{x}.$$

Then denoting $y = vx$ we obtain

$$(1 - v)xdx + vxdx + x^2dv = 0,$$

or

$$xdx + x^2dv = 0.$$

By integrating we have

$$x = e^{-v} + c,$$

or finally

$$x = e^{-y/x} + c$$

Exercise:

$$(x - 2y)dx + xdy = 0.$$

Solution:

It is easily seen that the differential equation is homogeneous. Then denoting $y = vx$ we obtain

$$(x - 2xv)dx + xvdx + x^2dv = 0.$$

That is

$$x(1 - v)dx + x^2dv = 0.$$

It is easily seen that an integrating factor is

$$\frac{1}{x^2(1-v)}.$$

Therefore,

$$\frac{dx}{x} = \frac{dv}{1-v}$$

By integrating we obtain

$$\log|x| = -\log|1 - v|,$$

or

$$x(1 - v) = c$$

Finally,

$$x - y = c.$$

Exercise:

$$\left(x^2 y + 2xy^2 - y^3 \right)dx - \left(2y^3 - xy^2 + x^3 \right)dy = 0.$$

Solution: The differential equation is homogeneous. Denote $y = vx$. Then

$$\left(x^3 v + 2x^3 v^2 - x^3 v^3 \right)dx - \left(2x^3 v^3 - x^3 v^2 + x^3 \right)(vdx + xdv) = 0,$$

$$\left(x^3 v + 2x^3 v^2 - x^3 v^3 \right)dx - \left(2x^3 v^4 - x^3 v^3 + x^3 v\right)dx - \left(2x^4 v^3 - x^4 v^2 + x^4 \right)dv = 0,$$

$$x^3 \left(2v^2 - 2v^4 \right)dx - x^4 \left(2v^3 - v^2 + 1\right)dv = 0,$$

$$\frac{dx}{x} = \frac{2v^3 - v^2 + 1}{2v^2 - 2v^4}dv$$

$$2 \log |x| = c_1 - \frac{1}{v} - \log \left|1 - v^2 \right|,$$

$$x^2 e^{-v} \left(1 - v^2 \right) = c,$$

$$c = \left(x^2 - y^2 \right)e^{x/y} .$$

Exercise:

$$\left(x \sin \frac{y}{x} - y \cos \frac{y}{x}\right) dx + x \cos \frac{y}{x} \, dy = 0.$$

Solution: It is readily seen that the differential equation is homogeneous. Putting $y = xv$ we obtain

$$\left(x \sin v - xv \cos v\right)dx + x \cos v\left(xdv + vdx\right) = 0,$$

$$\sin v dx + x \cos v dv = 0,$$

$$\frac{dx}{x} = - \tan v dv.$$

By integrating,

$$\log c|x| = \log \left| \cos v\right|,$$

$$cx = \cos v,$$

$$cx = \cos \frac{y}{x} .$$

Exercise:

$$\left(4x^4 - x^3y + y^4\right)dx + x^4dy = 0.$$

Solution: It is readily seen that the differential equation is homogeneous. Denote $y = vx$. Then,

$$\left(4x^4 - x^4v + x^4v^4\right)dx + x^4\left(xdv + vdx\right) = 0,$$

or

$$\left(4x^4 + x^4v^4\right)dx + x^5dv = 0,$$

$$\frac{dx}{x} = -\frac{dv}{4+v^4},$$

$$\log|cx| = -\int \frac{dv}{4+v^4},$$

$$cx = \exp\left(-\int \frac{dv}{4+v^4}\right).$$

Exercise:

$$\left(x^2 \sin \frac{y^2}{x^2} - 2y^2 \cos \frac{y^2}{x^2}\right)dx + 2xy \cos \frac{y^2}{x^2} dy = 0.$$

Solution: The differential equation is homogeneous. Denote $y = xv$. Then,

$$\left(x^2 \sin v^2 - 2x^2v^2 \cos v^2\right)dx + 2x^2v \cos v^2\left(xdv + vdx\right) = 0,$$

or

$$\sin v^2 dx + 2v \cos v^2\left(xdv\right) = 0,$$

$$\frac{dx}{x} = -2v \cot v^2 dv$$

$$\log|x| = -\int \frac{d\left(\sin v^2\right)}{\sin v^2} = -\log\left|\sin v^2\right| + \log c,$$

$$x \sin \frac{y^2}{x^2} = c,$$

Exercise:

$$(2x + y - 2)dx + \left(2y - x + 1\right)dy = 0.$$

Solution: The differential equation is not homogeneous. To reduce it to homogeneous, let us put $x = u + h, y = v + k$. Then,

$$(2u + 2h + v + k - 2)dx + (2v + 2k - u - h + 1)dy = 0.$$

Then we have the following system

$$2h + k = 2,$$

$$2k - h = -1.$$

We have $k = 0, h = 1$, and therefore,

$$(2u + v)du + (2v - u)dv = 0$$

is a homogeneous differential equation.

Exercise:

$$(x - y)dx + (x - y + 2)dy = 0.$$

Solution: The differential equation is not homogeneous. Putting $x - y = u$ we have

$$udu + 2(u + 1)dy = 0,$$

or

$$\frac{u}{2(u+1)}du = -dy$$

By integrating we obtain

$$y - c_1 = -\int \frac{u}{2(u+1)}du = \int \frac{1}{2(u+1)}du - \int \frac{u+1}{2(u+1)}du$$

$$= \frac{1}{2}\log|u + 1| - \frac{1}{2}(u + 1).$$

Finally,

$$x + y - \log|x + y - 1| = c.$$

Equations Reducible to Homogeneous Form

The differential equations under this category are of the form

$$\frac{dy}{dx} = \frac{a_1x + b_1y + c_1}{a_2x + b_2y + c_2}$$

while reducing this equation in homogeneous form there arise two cases.

Case I: When $\dfrac{a_1}{a_2} \neq \dfrac{b_1}{b_2}$,

we substitute $x = X + h$, $y = Y + k$ hence $\dfrac{dy}{dx} = \dfrac{dY}{dX}$, $\dfrac{dy}{dx} = \dfrac{a_1 x + b_1 y + c_1}{a_2 x + b_2 y + c_2}$ becomes

$$\frac{dY}{dX} = \frac{a_1(X+h) + b_1(Y+k) + c_1}{a_2(X+h) + b_2(Y+k) + c_2} = \frac{(a_1 X + b_1 Y) + (a_1 h + b_1 k + c_1)}{(a_2 X + b_2 Y) + (a_2 h + b_2 k + c_2)}$$

where h and k are some constants choosen such that

$$a_1 h + b_1 k + c_1 = 0,\ a_2 h + b_2 k + c_2 = 0$$

Then equation $\dfrac{dY}{dX} = \dfrac{a_1(X+h) + b_1(Y+k) + c_1}{a_2(X+h) + b_2(Y+k) + c_2} = \dfrac{(a_1 X + b_1 Y) + (a_1 h + b_1 k + c_1)}{(a_2 X + b_2 Y) + (a_2 h + b_2 k + c_2)}$ reduces in the form

$$\frac{dY}{dX} = \frac{a_1 X + b_1 Y}{a_2 X + b_2 Y}$$

which is a homogeneous. The values of h, k obtained from above two equations

Case II: If $\dfrac{a_1}{a_2} = \dfrac{b_1}{b_2}$, in this case we cannot calculate the value of the constants h and k. So, we have

$$\frac{a_1}{a_2} = \frac{b_1}{b_2} = k \Rightarrow a_1 = a_2 k \text{ and } b_1 = b_2 k$$

Hence the equation $\dfrac{dy}{dx} = \dfrac{a_1 x + b_1 y + c_1}{a_2 x + b_2 y + c_2}$ becomes $\dfrac{dy}{dx} = \dfrac{(a_1 x + b_1 y) + c_1}{k(a_1 x + b_1 y) + c_2}$

Here we substitute $a_1 x + b_1 y = z$ and it can be solve by variable separation method.

Example: Solve $\dfrac{dy}{dx} = \dfrac{3x - 2y + 1}{6x - 4y + 1}$.

Solution: $\dfrac{a_1}{a_2} = \dfrac{3}{6} = \dfrac{1}{2}$ and $\dfrac{b_1}{b_2} = \dfrac{2}{4} = \dfrac{1}{2} \Rightarrow \dfrac{a_1}{a_2} = \dfrac{b_1}{b_2}$

The given equation may be written as

$$\frac{dy}{dx} = \frac{3x - 2y + 1}{2(3x - 2y) + 1}$$

Let $3x - 2y = z \Rightarrow 3 - 2\dfrac{dy}{dx} = \dfrac{dz}{dx} \Rightarrow \dfrac{dy}{dx} = \dfrac{1}{2}\left(3 - \dfrac{dz}{dx}\right)$

Using this value in equation $\dfrac{dy}{dx} = \dfrac{3x - 2y + 1}{2(3x - 2y) + 1}$, we get

$$\frac{1}{2}\left(3-\frac{dz}{dx}\right)=\frac{z+1}{2z+1}\Rightarrow\frac{dz}{dx}=3-\frac{2z+2}{2z+1}=\frac{4z+1}{2z+1}$$

$$\text{or }\frac{2z+1}{4z+1}dz=dx\Rightarrow\int\frac{2z+1}{4z+1}dz=\int dx+C$$

$$\Rightarrow\int\left[\frac{1}{2}+\frac{1}{2(4z+1)}\right]dz=x+C\Rightarrow\frac{z}{2}+\frac{1}{8}\log(4z+1)=x+C$$

$$\Rightarrow z+\frac{1}{4}\log4+\frac{1}{4}\log\left(z+\frac{1}{4}\right)=2x+2C$$

Now, replace z by $3x-2y$, we get

$$\frac{1}{4}\log\left(3x-2y+\frac{1}{4}\right)=2x-(3x-2y)+C_1$$

$$\text{or }4x-8y+\log\left(3x-2y+\frac{1}{4}\right)=C_1'.\quad\text{Ans.}$$

Example: Solve

$$\frac{dy}{dx}=\frac{y-x+1}{y+x+5}.$$

Solution:

$$\frac{dy}{dx}=\frac{y-x+1}{y+x+5}$$

$$\frac{a_1}{a_2}=-1,\frac{b_1}{b_2}=1,\text{ so let }x=X+h\text{ and }y=Y+k$$

Then the equation $\dfrac{dy}{dx}=\dfrac{y-x+1}{y+x+5}$, reduces in the form

$$\frac{dY}{dX}=\frac{(Y+k)-(X+h)+1}{(Y+k)+(X+h)+5}=\frac{(Y-X)+(k-h+1)}{(Y+X)+(k+h+5)}$$

Let us choose h and k such that

$$k-h+1=0,\;k+h+5=0$$

which give $h=-2,\;k=-3$, then $\dfrac{dY}{dX}=\dfrac{(Y+k)-(X+h)+1}{(Y+k)+(X+h)+5}=\dfrac{(Y-X)+(k-h+1)}{(Y+X)+(k+h+5)}$ becomes

$\dfrac{dY}{dX}=\dfrac{Y-X}{Y+X}$ which is homogeneous.

Putting $Y=vX$, we get

$$v + X\frac{dv}{dX} = \frac{X(v-1)}{X(v+1)} \Rightarrow X\frac{dv}{dX} = \frac{v-1}{v+1} - v$$

$$\Rightarrow X\frac{dv}{dX} = \frac{v-1-v^2-v}{(v+1)} = -\frac{v^2+1}{v+1}$$

$$\Rightarrow \left(\frac{v+1}{v^2+1}\right)dv = -\frac{dX}{X}, \text{ on integrating}$$

$$\int\left(\frac{v}{v^2+1} + \frac{1}{v^2+1}\right)dv = -\int\frac{dx}{x} + \log C$$

$$\Rightarrow \frac{1}{2}\log(v^2+1) + \tan^{-1}v = -\log X + \log C$$

$$\Rightarrow \tan^{-1}\frac{Y}{X} + \log\sqrt{v^2+1} + \log X - \log C = 0$$

$$\Rightarrow \tan^{-1}\frac{Y}{X} + \log\sqrt{Y^2+X^2} - \log C = 0$$

$$\Rightarrow \tan^{-1}\frac{y+3}{x+2} + \log\sqrt{(x+2)^2+(y+3)^2} - \log C = 0 \qquad | \text{As } X = x+2, Y = y+3$$

$$\Rightarrow \tan^{-1}\frac{y+3}{x+2} + \log\left\{\frac{1}{C}\sqrt{(x+2)^2+(y+3)^2}\right\} = 0. \quad \text{Ans.}$$

Exact Differential Equations

A differential equation of type

$$P(x,y)dx + Q(x,y)dy = 0$$

is called an exact differential equation if there exists a function of two variables $u(x,y)$ with continuous partial derivatives such that

$$du(x,y) = P(x,y)dx + Q(x,y)dy.$$

The general solution of an exact equation is given by

$$u(x,y) = C,$$

where C is an arbitrary constant.

Test for Exactness

Let functions $P(x,y)$ and $Q(x,y)$ have continuous partial derivatives in a certain domain D. The differential equation $P(x,y)dx + Q(x,y)dy = 0$ is an exact equation if and only if

$$\frac{\partial Q}{\partial x} = \frac{\partial P}{\partial y}.$$

Algorithm for Solving an Exact Differential Equation

Step 1: First it's necessary to make sure that the differential equation is *exact* using the *test for exactness*,

$$\frac{\partial Q}{\partial x} = \frac{\partial P}{\partial y}.$$

Step 2: Then we write the system of two differential equations that define the function $u(x,y)$,

$$\begin{cases} \dfrac{\partial u}{\partial x} = P(x,y) \\ \dfrac{\partial u}{\partial y} = Q(x,y) \end{cases}.$$

Step 3: Integrate the first equation over the variable x. Instead of the constant C, we write an unknown function of y,

$$u(x,y) = \int P(x,y)dx + \varphi(y).$$

Step 4: Differentiating with respect to y, we substitute the function $u(x,y)$ into the second equation,

$$\frac{\partial u}{\partial y} = \frac{\partial}{\partial y}\left[\int P(x,y)dx + \varphi(y)\right] = Q(x,y).$$

From here we get expression for the derivative of the unknown function $\varphi(y)$:

$$\varphi'(y) = Q(x,y) - \frac{\partial}{\partial y}\left(\int P(x,y)dx\right).$$

Step 5: By integrating the last expression, we find the function $\varphi(y)$ and, hence, the function $u(x,y)$,

$$u(x,y) = \int P(x,y)dx + \varphi(y).$$

Step 6: The general solution of the exact differential equation is given by

$$u(x,y) = C.$$

In Step 3, we can integrate the second equation over the variable y instead of integrating the first equation over x. After integration we need to find the unknown function $\psi(x)$.

Example: Solve the differential equation $\left(2xy - 3x^2\right)dx + \left(x^2 - 2y\right)dy = 0$.

Solution: The given differential equation is exact because

$$\frac{\partial M}{\partial y} = \frac{\partial}{\partial y}[2xy - 3x^2] = 2x = \frac{\partial N}{\partial x} = \frac{\partial}{\partial x}[x^2 - 2y].$$

The general solution $f(x, y) = C$, is given by

$$f(x, y) = \int M(x, y)\, dx$$

$$= \int \left(2xy - 3x^2\right) dx = x^2 y - x^3 + g(y)$$

We can determine $g(y)$ by integrating $N(x, y)$ with respect to y and reconciling the two expressions for $f(x, y)$. An alternative method is to partially differentiate this version of $f(x, y)$ with respect to y and compare the result with $N(x, y)$. In other words,

$$f_y(x, y) = \frac{\partial}{\partial y}[x^2 y - x^3 + g(y)] = x^2 + g'(y) = \overbrace{x^2 - 2y}^{N(x,y)}.$$

$$\boxed{g(y) = -2y}$$

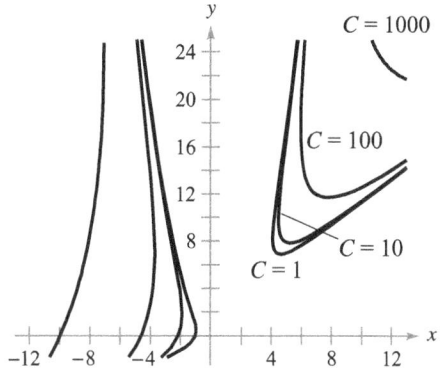

Thus, $g'(y) = -2y$ and it follows that $g(y) = -y^2 + C_1$. Therefore, $f(x, y) = x^2 y - x^3 - y^2 + C_1$

and the general solution is $x^2 y - x^3 - y^2 + C$. Above figure shows the solution curves that correspond to $C = 1,\ 10,\ 100,\ \text{and}\ 1000$.

Example: Find the particular solution of

$$\left(\cos x - x \sin x + y^2\right) dx + 2xy\, dy = 0$$

that satisfies the initial condition $y = 1$ when $x = \pi$

Solution: The differential equation is exact because

$$\overbrace{\frac{\partial}{\partial y}[\cos x - x \sin x + y^2]}^{\frac{\partial M}{\partial y}} = 2y = \overbrace{\frac{\partial}{\partial x}[2xy]}^{\frac{\partial N}{\partial x}}.$$

Because $N(x,y)$ is simpler than $M(x,y)$, it is better to begin by integrating $N(x,y)$

$$f(x,y) = \int N(x,y)\, dy = \int 2xy\, dy = xy^2 + g(x)$$

$$f_x(x,y) = \frac{\partial}{\partial x}[xy^2 + g(x)] = y^2 + g'(x) = \overbrace{\cos x - x \sin x + y^2}^{M(x,y)}$$

$$\boxed{g'(x) = \cos x - x \sin x}$$

Thus, $g'(x) = \cos x - x \sin x$ and

$$g(x) = \int (\cos x - x \sin x)\, dx$$

$$= x \cos x + C_1$$

which implies that $f(x,y) = xy^2 + x \cos x + C_1$, and the general solution is

$$xy^2 + x \cos x = C. \qquad \text{General solution}$$

Applying the given initial condition produces

$$\pi(1)^2 + \pi \cos \pi = C$$

which implies that $C = 0$ Hence, the particular solution is

$$xy^2 + x \cos x = 0. \qquad \text{Particular solution}$$

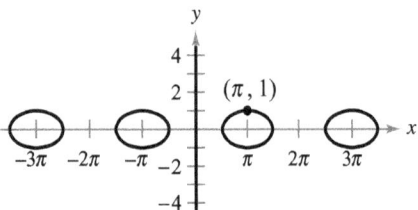

The graph of the particular solution is shown in above Figure. Notice that the graph consists of two parts: the ovals are given by $y^2 + \cos x = 0$, and the y-axis is given by $x = 0$.

In above Example, note that If $z = f(x,y) = xy^2 + x \cos x$, the total differential of z is given by

$$dz = f_x(x,y)\, dx + f_y(x,y)\, dy$$

$$= (\cos x - x \sin x + y^2)\, dx + 2xy\, dy$$

$$= M(x,y)\, dx + N(x,y)\, dy.$$

In other words, $M\,dx + N\,dy = 0$ is called an exact differential equation because $M\,dx + N\,dy$ is exactly the differential of $f(x,y)$.

Example: Solve the differential equation $\dfrac{dy}{dx} = -\dfrac{x+4y}{4x-y}$.

Solution: $\dfrac{dy}{dx} = -\dfrac{x+4y}{4x-y}$

$$(x+4y)\,dx + (4x-y)\,dy = 0$$

Here, $P(x,y) = x+4y$ and $Q(x,y) = 4x-y$

Let us first test for exactness

$$\frac{\partial P}{\partial y} = 4$$

$$\frac{\partial Q}{\partial x} = 4$$

Since $\dfrac{\partial P}{\partial y} = \dfrac{\partial Q}{\partial y}$

Hence given equation is an exact differential equation.

Now,

$$f(x,y) = \int P\partial x$$

$$= \int (x+4y)\partial x$$

$$= \frac{x^2}{2} + 4xy + c_1$$

and

$$f(x,y) = \int Q\partial y$$

$$= \int (4x - y)\partial y$$

$$= 4xy - \frac{y^2}{2} + c_2$$

Merging these two expressions, we get

$$f(x,y) = \frac{x^2}{2} + 4xy - \frac{y^2}{2}$$

So, the solution will be

$$\frac{x^2}{2} + 4xy - \frac{y^2}{2} = C$$

$$x^2 + 8xy - y^2 = C$$

The examples of solving exact differential equations are as follows.

Example: Solve the differential equation $(2xy - 9x^2)dx + (2y + x^2 + 1)dy = 0$ at $y(0) = 3$.

Solution: $(2xy - 9x^2)dx + (2y + x^2 + 1)dy = 0$

Here, $P(x,y) = (2xy - 9x^2)$ and $Q(x,y) = (2y + x^2 + 1)$

Let us test for exactness

$$\frac{\partial P}{\partial y} = 2x$$

$$\frac{\partial Q}{\partial x} = 2x$$

Since, $\dfrac{\partial P}{\partial y} = \dfrac{\partial Q}{\partial x}$

Hence, the given equation is an exact differential equation.

Now,

$$f(x,y) = \int P \partial x$$

$$= \int (2xy - 9x^2)\partial x$$

$$= x^2 y - 3x^3 + c_1$$

and

$$f(x,y) = \int Q \partial y$$

$$= \int (2y + x^2 + 1)\partial y$$

$$= y^2 + x^2 y + y + c_2$$

Merging these two expressions, we get

$$f(x,y) = x^2y - 3x^3 + y^3 + y$$

So, the solution will be

$$x^2y - 3x^3 + y^2 + y = C$$

In order to find solution at y(0) = 3, we substitute x = 0 and y = 3 in the above solution $0 - 0 + 3^2 + 3 = C$

$$C = 12$$

So, $x^2y - 3x^3 + y^2 + y = 12$ is the required solution.

Example: Evaluate the solution for following exact differential equation:

$$\left(y^2 - 2x\right)dx + \left(2xy + 1\right)dy = 0$$

Solution: $\left(y^2 - 2x\right)dx + \left(2xy + 1\right)dy = 0$

Here, $P(x,y) = \left(y^2 - 2x\right)$ and $Q(x,y) = \left(2\,xy + 1\right)$

Let us test for the exactness

$$\frac{\partial P}{\partial y} = 2y$$

$$\frac{\partial Q}{\partial x} = 2y$$

Here, $\dfrac{\partial P}{\partial y} = \dfrac{\partial Q}{\partial x}$

so, the given equation is an exact differential equation.

Now,

$$f(x,y) = \int P\partial x$$

$$= \int (y^2 - 2x)\partial x$$

$$= xy^2 - x^2 + c_1$$

and

$$f(x,y) = \int Q\partial y$$

$$= \int (2xy + 1)\partial y$$

$$= xy^2 + y + c_2$$

Merging these two expressions, we get

$$f(x,y) = xy^2 - x^2 + y$$

Hence, the solution will be

$$xy^2 - x^2 + y = C$$

Integrating Factors

Let p and g be functions of t and consider the following first order differential equation:

$$\frac{dy}{dt} + p(t)y = g(t)$$

If we multiply the both sides of the equation above by the function $\mu(t)$ we get that:

$$\mu(t)\frac{dy}{dt} + \mu(t)p(t)y = \mu(t)g(t)$$

If we can guarantee that $\mu'(t) = \mu(t)p(t)$, then notice that by applying the product rule for differentiation that we get: $\frac{d}{dt}(\mu(t)y) = \mu(t)\frac{dy}{dt} + \mu'(t)y$ and substituting $\mu'(t) = \mu(t)y$ and we get that $\frac{d}{dt}(\mu(t)y) = \mu(t)\frac{dy}{dt} + \mu(t)p(t)y$ which is exactly the left hand side of the equation above. Thus we get that:

$$\frac{d}{dt}(\mu(t)y) = \mu(t)g(t)$$

The above differential equation can be solved by integrating both sides of the equation with respect to t and isolating y. The question now arises on how we can find such a function

$$\mu(t).$$

If $\frac{dy}{dt} + p(t)y = g(t)$ is a first order differential equation, then $\mu(t)$ is called an Integrating Factor if for $\mu(t)\frac{dy}{dt} + \mu(t)p(t)y = \mu(t)g(t)$ we have that $\mu'(t) = \mu(t)p(t)$.

The following proposition will give us a formula for obtaining the integrating factor for differential equations in the form $\frac{dy}{dt} + p(t)y = g(t)$.

Proposition: If $\frac{dy}{dt} + p(t)y = g(t)$ is a differential equation, then an integrating factor $\mu(t)$ of this equation is given by the formula $\mu(t) = e^{\int p(t)\,dt}$.

Proof: We want to find $\mu(t)$ such that $\mu'(t) = \mu(t)p(t)$. We can rewrite this equation as as $\dfrac{\mu'(t)}{\mu(t)} = p(t)$ and then:

$$\frac{\mu'(t)}{\mu(t)} = p(t)$$

$$\frac{d}{dt} \ln |\mu(t)| = p(t)$$

$$\int \frac{d}{dt} \ln |\mu(t)| \, dt = \int p(t) \, dt$$

$$\ln |\mu(t)| = \int p(t) \, dt$$

$$\mu(t) = \pm e^{\int p(t) \, dt}$$

Since we only need one integrating factor to solve differential equations in the form $\dfrac{dy}{dt} + p(t)y = g(t)$, we can more generally note that $\mu(t) = e^{\int p(t) \, dt}$ is an integrating factor of this differential equation.

The above proposition, integrating factors $\mu(t)$ are not unique. In fact, there are infinitely many integrating factors. This can be see when evaluating the indefinite integral in $\mu(t) = e^{\int p(t) \, dt}$ which will result in getting $\mu(t) = e^{P(t)+C}$ where P is any antiderivative if p and where C is a constant. We will always use the simplest integrating factor in solving differential equations of this type.

Let's now look at some examples of applying the method of integrating factors.

Example: Find all solutions to the differential equation $\dfrac{dy}{dt} + \dfrac{2y}{t} = \dfrac{\sin t}{t^2}$.

We first notice that our differential equation is in the appropriate form $\dfrac{dy}{dt} + p(t)y = g(t)$ where $p(t) = \dfrac{2}{t}$ and $g(t) = \dfrac{\sin t}{t^2}$. We compute our integrating factor as:

$$\mu(t) = e^{\int p(t) \, dt} = e^{\int \frac{2}{t} \, dt} = e^{2\ln(t)} = e^{\ln(t^2)} = t^2$$

Thus we have that for C as a constant:

$$\mu(t)\frac{dy}{dt} + \mu(t)p(t)y = \mu(t)g(t)$$

$$t^2 \frac{dy}{dt} + 2ty = \sin t$$

$$\frac{d}{dt}\left(t^2 y\right) = \sin t$$

$$\int \frac{\partial}{\partial t}\left(t^2 y\right) \, dt = \int \sin t \, dt$$

$$t^2 y = -\cos t + C$$

$$y = \frac{-\cos t}{t^2} + \frac{C}{t^2}$$

Example: Find all solutions to the differential equation $\frac{dy}{dt} - \frac{y}{t} + te^{-t} = 0$.

We first rewrite our differential equation as $\frac{dy}{dt} - \frac{y}{t} = -te^{-t}$. We note that in this form we have $p(t) = -\frac{1}{t}$ and $g(t) = -te^{-t}$. We now find an integrating factor:

$$\mu(t) = e^{\int p(t)\, dt} = e^{\int -\frac{1}{t}\, dt} = e^{-\ln t} = e^{\ln\left(\frac{1}{t}\right)} = \frac{1}{t}$$

Thus we have that for C as a constant:

$$\mu(t)\frac{dy}{dt} + \mu(t)p(t)y = \mu(t)g(t)$$

$$\frac{1}{t}\frac{dy}{dt} - \frac{1}{t^2}y = -e^{-t}$$

$$\frac{dy}{dt}\left(\frac{y}{t}\right) = -e^{-t}$$

$$\int \frac{dy}{dt}\left(\frac{y}{t}\right) dt = -\int e^{-t}\, dt$$

$$\frac{y}{t} = e^{-t} + C$$

$$y = te^{-t} + tC$$

Two special cases will be considered:

- Case 1:

  Consider the differential equation $M\, dx + N\, dy = 0$. If this equation is not exact, then M_y will not equal N_x; that is, $M_y - N_x \neq 0$. However, if

  $$\frac{M_y - N_x}{N}$$

  is a function of x only, let it be denoted by $\xi(x)$. Then

  $$\mu(x) = e^{\int \xi(x)\, dx}$$

  will be an integrating factor of the given differential equation.

- Case 2:

  Consider the differential equation $M\, dx + N\, dy = 0$. If this equation is not exact, then M_y will not equal N_x; that is, $M_y - N_x \neq 0. = 0$. However, if

  $$\frac{M_y - N_x}{-M}$$

is a function of y only, let it be denoted by $\psi(y)$. Then

$$\mu(y) = e^{\int \psi(y)\,dy}$$

will be an integrating factor of the given differential equation.

Example: The equation

$$(3xy - y^2)\,dx + x(x - y)\,dy = 0$$

is not exact, since

$$M_y = \frac{\partial}{\partial y}(3xy - y^2) = 3x - 2y \text{ but } N_x = \frac{\partial}{\partial x}(x^2 - xy) = 2x - y$$

However, note that

$$\frac{M_y - N_x}{N} = \frac{(3x - 2y) - (2x - y)}{x(x - y)} = \frac{x - y}{x(x - y)} = \frac{1}{x}$$

is a function of x alone. Therefore, by Case 1,

$$e^{\int (1/x)\,dx} = e^{\ln x} = x$$

will be an integrating factor of the differential equation. Multiplying both sides of the given equation by $\mu = x$ yields

$$\underbrace{(3x^2 y - xy^2)}_{\mu M = \overline{M}}\,dx + \underbrace{(x^3 - x^2 y)}_{\mu N = \overline{N}}\,dy = 0$$

which is exact because

$$\frac{\partial \overline{M}}{\partial y} = 3x^2 - 2xy = \frac{\partial \overline{N}}{\partial x}$$

Solving this equivalent exact equation by the method, M is integrated with respect to x,

$$\int \overline{M}\,\partial x = \int (3x^2 y - xy^2)\,\partial x = x^3 y - \frac{1}{2}x^2 y^2$$

and N integrated with respect to y:

$$\int \overline{N}\,\partial y = \int (x^3 - x^2 y)\,\partial y = x^3 y - \frac{1}{2}x^2 y^2$$

(with each "constant" of integration ignored, as usual). These calculations clearly give

$$x^3 y - \frac{1}{2}x^2 y^2 = c$$

as the general solution of the differential equation.

Example: The equation

$$(x+y)\sin y\, dx + (x\sin y + \cos y)\, dy = 0$$

is not exact, since

$$M_y = (x+y)\,\cos y + \sin y \text{ but } N_x = \sin y$$

However, note that

$$\frac{M_y - N_x}{-M} = \frac{(x+y)\,\cos y + \sin y - \sin y}{-(x+y)\sin y} = -\frac{\cos y}{\sin y}$$

is a function of y alone (Case 2). Denote this function by $\psi(y)$; since

$$\int \psi(y)\, dy = -\int \frac{\cos y\, dy}{\sin y} = -\ln(\sin y)$$

the given differential equation will have

$$e^{\int \psi(y)\, dy} = e^{-\ln(\sin y)} = e^{\ln(\sin y)^{-1}} = (\sin y)^{-1}$$

as an integrating factor. Multiplying the differential equation through by $\mu = (\sin y)^{-1}$ yields

$$\underbrace{(x+y)\, dx}_{\mu M\,=\,\overline{M}} + \underbrace{\left(x + \frac{\cos y}{\sin y}\right)\, dy}_{\mu N\,=\,\overline{N}} = 0$$

which *is* exact because

$$\overline{M}_y = 1 = \overline{N}_x$$

To solve this exact equation, integrate M with respect to x and integrate N with respect to y, ignoring the "constant" of integration in each case:

$$\int \overline{M}\, \partial x = \int (x+y)\, \partial x = \frac{1}{2}x^2 + xy$$

$$\int \overline{N}\, \partial y = \int \left(x + \frac{\cos y}{\sin y}\right)\, \partial y = xy + \ln |\sin y|$$

These integrations imply that

$$\frac{1}{2}x^2 + xy + \ln |\sin y| = c$$

is the general solution of the differential equation.

Example: Solve the IVP

$$\left(3e^x y + x\right) dx + e^x \, dy = 0$$
$$y(0) = 1$$

The given differential equation is not exact, since

$$M_y = \frac{\partial}{\partial y}\left(3e^x y + x\right) = 3e^x \text{ but } N_x = \frac{\partial}{\partial x}\left(e^x\right) = e^x$$

However, note that

$$\frac{M_y - N_x}{N} = \frac{3e^x - e^x}{e^x} = 2$$

which can be interpreted to be, say, a function of x only; that is, this last equation can be written as $\xi(x) \equiv 2$. Case 1 then says that

$$e^{\int \xi(x)\,dx} = e^{\int 2\,dx} = e^{2x}$$

will be an integrating factor. Multiplying both sides of the differential equation by $\mu(x) = e^{2x}$ yields

$$\underbrace{\left(3e^{3x} y + xe^{2x}\right)}_{\mu M \,=\, \overline{M}} dx + \underbrace{\left(e^{3x}\right)}_{\mu N \,=\, \overline{N}} dy = 0$$

which *is* exact because

$$\overline{M}_y = 3e^{3x} = \overline{N}_x$$

Now, since

$$\int \overline{M}\,\partial x = \int\left(3e^{3x} y + xe^{2x}\right) \partial x = e^{3x} y + \frac{1}{2} xe^{2x} - \frac{1}{4}e^{2x}$$

And

$$\int \overline{N}\,\partial y = \int e^{3x}\,\partial y = e^{3x} y$$

(with the "constant" of integration suppressed in each calculation), the general solution of the differential equation is

$$e^{3x} y + \frac{1}{2} xe^{2x} - \frac{1}{4}e^{2x} = c$$

The value of the constant c is now determined by applying the initial condition $y(0) = 1$:

$$\left[e^{3x}y + \frac{1}{2}xe^{2x} - \frac{1}{4}e^{2x} \right]_{x=0, \, y=1} = c \Rightarrow \frac{3}{4} = c$$

Thus, the particular solution is

$$e^{3x}y + \frac{1}{2}xe^{2x} - \frac{1}{4}e^{2x} = \frac{3}{4}$$

which can be expressed explicitly as

$$y = \frac{3e^{-3x} + e^{-x}(1-2x)}{4}$$

Example: Given that the nonexact differential equation

$$\left(5xy^2 - 2y \right)dx + \left(3x^2y - x \right)dy = 0$$

has an integrating factor of the form $\mu(x,y) = x^a y^b$ for some positive integers a and b, find the general solution of the equation.

Since there exist positive integers a and b such that $x^a y^b$ is an integrating factor, multiplying the differential equation through by this expression must yield an exact equation. That is,

$$\underbrace{\left(5x^{a+1}y^{b+2} - 2x^a y^{b+1} \right)}_{\mu M = \overline{M}} dx + \underbrace{\left(3x^{a+2}y^{b+1} - x^{a+1}y^b \right)}_{\mu N = \overline{N}} dy = 0$$

is exact for some a and b. Exactness of this equation means

$$\overline{M}_y = \overline{N}_x$$

$$5(b+2)x^{a+1}y^{b+1} - 2(b+1)x^a y^b = 3(a+2)x^{a+1}y^{b+1} - (a+1)x^a y^b$$

By equating like terms in this last equation, it must be the case that

$$5(b+2) = 3(a+2) \text{ and } 2(b+1) = a+1$$

The simultaneous solution of these equations is $a = 3$ and $b=1$.

Thus the integrating factor $x^a y^b$ is $x^3 y$, and the exact equation $M\,dx + N\,dy = 0$ reads

$$\left(5x^4 y^3 - 2x^3 y^2 \right)dx + \left(3x^5 y^2 - x^4 y \right)dy = 0$$

Now, since

$$\int \overline{M} \partial x = \int \left(5x^4 y^3 - 2x^3 y^2 \right)dx = x^5 y^3 - \frac{1}{2}x^4 y^2$$

and

$$\int \overline{N} \partial y = \int \left(3x^5 y^2 - x^4 y\right) \partial y = x^5 y^3 - \frac{1}{2}x^4 y^2$$

(ignoring the "constant" of integration in each case), the general solution of the differential equation $\underbrace{\left(5x^{a+1}y^{b+2} - 2x^a y^{b+1}\right)}_{\mu M = \overline{M}} dx + \underbrace{\left(3x^{a+2}y^{b+1} - x^{a+1}y^b\right)}_{\mu N = \overline{N}} dy = 0$ and hence the original differential equation—is clearly

$$x^5 y^3 - \frac{1}{2}x^4 y^2 = c$$

Backwards Product Rule

The next step of the Integrating Factor Technique is an iffy one, because it requires careful observation. You must be able to condense a sum of terms into the derivative of a product. Luckily, since these terms will usually have exponential factors (due to the integrating factor introducing exponential factors into the equation), the product rule application is usually easy to spot.

Example: Write $e^{x^2} + 2x^2 e^{x^2}$ as the derivative of a product.

Solution: Notice that there is an exponential factor in both terms. Upon differentiation of an exponential term, the argument of the exponent will not change. This means the sum can be written as

$$(f(x)e^{x^2})'$$

While observation is usually sufficient to find f(x), we will show a more thorough way to find f(x). Differentiate the above expression with the Product Rule and Chain Rule, then set it equal to the original expression:

$$f'(x)e^{x^2} + 2xf(x)e^{x^2} = e^{x^2} + 2x^2 e^{x^2}$$

Divide both sides of the equation by e^{x^2} (which never equals zero):

$$f'(x) + 2xf(x) = 1 + 2x^2$$

From here it is obvious that the left-hand side must contain a constant term and a quadratic term. f(x) must be linear to make this true.

$$2xf(x) = 2x^2$$
$$\Rightarrow f(x) = x$$
$$\Rightarrow f'(x) = 1$$

This makes the equation true, so the condensed form of the original expression is $\left(xe^{x^2}\right)'$. Feel free to check this by expanding and differentiating.

Example: Write the expression $x^5 e^x + 5x^4 e^x$ as the derivative of a product.

Solution: Here we will show a different method. See that each term in the sum has two factors that can be paired up: x^5 with $5x^4$, and e^x with e^x. Each pair of factors differs only by one differentiation, which suggests, by the Product Rule, that

$$(x^5 e^x)' = x^5 e^x + 5x^4 e^x$$

Differentiating the left-hand side confirms that this is true.

Example: Write the expression $2e^{2x} \sin x + e^{2x} \cos x$ as the derivative of a product of two factors.

Solution: Both terms have the same exponential factor, that must be one of the factors, so

$$(f(x)e^{2x})' = 2e^{2x} \sin x + e^{2x} \cos x$$

Expand the left side with the Product Rule and the Chain Rule:

$$f'(x)e^{2x} + 2f(x)e^{2x} = 2e^{2x} \sin x + e^{2x} \cos x$$

The second term on the left side must be equivalent to one of the two terms on the right side. If

$$2f(x)e^{2x} = 2e^{2x} \sin x$$

then $f(x) = \sin x$ and $f'(x) = \cos x$. Substituting gives

$$\cos x e^{2x} + 2\sin x e^{2x} = 2e^{2x} \sin x + e^{2x} \cos x$$

That means that $f(x) = \sin x$ and the desired form of the expression is $(e^{2x} \sin x)'$.

Alternate Solution: Pair up the factors between the two terms as $2e^{2x}$ with e^{2x} and $\sin x$ with $\cos x$. This is a logical pairing because the two pieces to each pair differ by one differentiation. Differentiating e^{2x} gives $2e^{2x}$ and differentiating $\sin(x)$ gives $\cos(x)$. Therefore, in the Product Rule we let $f(x) = e^{2x}$, $f'(x) = 2e^{2x}$, $g(x) = \sin(x)$, and $g'(x) = \cos(x)$, implying that the desired form of the sum is $(e^{2x} \sin(x))'$.

Integrating the Derivative

The last major step in the Integrating Factor Technique is to integrate the equation. At this point one side of the equation will contain a derivative of a product. Integrating this side of the equation will result in the argument of the derivative plus a constant. The other side of the equation should be pretty easy to integrate.

Example: Integrate and solve for y:

$$(y e^x \tan x)' = 2 \sin x$$

Solution: Integrate both sides:

$$\int (ye^x \tan x)' dx = \int 2 \sin x dx$$

$$\Rightarrow ye^x \tan x + C_1 = -2 \cos x + C_2$$

Move the constants to one side, and rename $C_2 - C_1$ as C:

$$ye^x \tan x = -2 \cos x + C_2$$

Now isolate y:

$$y = \frac{-2 \cos x + C}{e^x \tan x}$$

Combining the Steps

Now we will put the steps together to get the entire Integrating Factor Technique. Here are the steps for reference:

1. Ensure that the differential equation is of the appropriate type (linear and first-order) and that it is of the form $y' + f(x)y = g(x)$.

2. Determine the integrating factor and multiply both sides of the differential equation by it.

3. Write one side of the differential equation as the derivative of a product of two factors.

4. Solve for y by integrating the derivative.

Example: Solve the differential equation

$$y' + 6y = x$$

Solution: The equation is of the appropriate form. Use the y term to find $M(x)$:

$$M(x) = e^{\int 6dx} = e^{6x}$$

Multiply both sides of the differential equation by this function:

$$e^{6x} y' + 6e^{6x} y = xe^{6x}$$

For the derivative of a product, y will be one of the factors. Obviously, e^{6x} must be the other factor because it will not vanish upon differentiation $\left(\frac{d}{dx}(e^{6x}) = 6e^{6x}\right)$. Therefore,

$$(ye^{6x})' = xe^{6x}$$

Integrate:

$$\int (ye^{6x})' dx = \int xe^{6x} dx$$

The right side must be simplified with integration by parts:

$$\int xe^{6x}\,dx = \frac{1}{6}xe^{6x} - \int \frac{1}{6}e^{6x}dx = \frac{1}{6}xe^{6x} - \frac{1}{36}e^{6x} + C_1$$

That means

$$ye^{6x} + C_2 = \frac{1}{6}xe^{6x} - \frac{1}{36}e^{6x} + C_1$$

Isolate :

$$y = \frac{1}{6}x - \frac{1}{36} + C_1 e^{-6x}$$

Here is another example. However, this one is an initial value problem, meaning there is a specific value of C that needs to be found.

Example: Solve the following differential equation at the point $(1,4)$:

$$xy' + 2xy = x^3$$

Solution: The equation is not in the appropriate form. Divide both sides of the equation by x:

$$y' + 2y = x^2$$

Now find $M(x)$:

$$M(x) = e^{\int 2dx} = e^{2x}$$

Multiply both sides of the equation by $M(x)$:

$$y'e^{2x} + 2ye^{2x} = x^2$$

Obviously the two factors of the derivative that we need are y and e^{2x}:

$$(ye^{2x})' = x^2$$

Integrate:

$$\int (ye^{2x})'dx = \int x^2 dx \Rightarrow$$

$$ye^{2x} + C_1 = \frac{1}{3}x^3 + C_2$$

Solve for y:

$$ye^{2x} = \frac{1}{3}x^3 + C$$

$$\Rightarrow y = \frac{1}{3}x^3 e^{-2x} + Ce^{-2x}$$

From now on always assume that C represents the difference of the two constants that are formed upon integration of the differential equation. Now we must consider the initial value (1, 4). Plug these values in for x and y to solve for C :

$$4 = \frac{1}{3}(1)^3 e^{-2(1)} + Ce^{-2(1)}$$

$$\Rightarrow 4 = \frac{1}{3}(1)e^{-2} + Ce^2$$

$$\Rightarrow 4 - \frac{e^{-2}}{3} = Ce^{-2}$$

$$\Rightarrow 4e^2 - \frac{1}{3} = C$$

$$C = \frac{12e^2 - 1}{3}$$

This means the solution to the differential equation is

$$y = \frac{1}{3}x^3 e^{-2x} + \frac{e^{-2x}\left(12e^2 - 1\right)}{3}$$

Example: Solve the differential equation

$$(y' - 5x^4)^3 = 27x^6 y^3$$

Solution: It will take some work to get this differential equation into the appropriate form. Take the cube root of both sides of the equation:

$$y' - 5x^4 = 3x^2 y$$

Now move the x term to the right side of the equation:

$$y' = 3x^2 y + 5x^4$$

Finally, move the term with the y to the left side of the equation:

$$y' - 3x^2 y = 5x^4$$

The equation is now in the appropriate form. Find $M(x)$:

$$M(x) = e^{\int -3x^2 dx} = e^{-x^3}$$

Multiply both sides of the equation by $M(x)$:

$$y'e^{-x^3} - 3x^2 ye^{-x^3} = 5x^4 e^{-x^3}$$

Write the left side as the derivative of a product:

$$(ye^{-x^3})' = 5x^4 e^{-x^3}$$

Integrate:

$$\int (ye^{-x^3})' dx = \int 5x^4 e^{-x^3} dx$$

The left side is easy to integrate due to the derivative in the integrand.

Bernoulli's Equation

A differential equation in the form $y' + p(x)y = g(x)y^n$ is called a Bernoulli differential equation. These differential equations are not linear, however, we can "convert" them to be linear.

We first let $v = y^{1-n}$. Then $v' = (1-n)y^{-n}y'$. We then take the differential equation above and divide both sides of it by y^n and apply these substitutions to get:

$$y' + p(x)y = g(x)y^n$$

$$y^{-n}y' + p(x)y^{1-n} = g(x)$$

$$\left(\frac{1}{1-n}\right)v' + p(x)v = g(x)$$

We then solve this differential equation as a first order linear equation for v, and subsequently solve y from v.

Example Solve the differential equation $6y' - 2y = ty^4$.

Solution: It's not hard to see that this is indeed a Bernoulli differential equation. We first divide by 6 to get this differential equation in the appropriate form:

$$y' - \frac{1}{3}y = \frac{t}{6}y^4$$

In this case, we have that $n = 4$. Let $y^{1-4} = y^{-3}$. Then $v' = -3y^{-4}y'$ and $-\frac{v'}{3} = y^{-4}y'$. Now we divide both sides of the differential equation above by y^4 and apply these substitutions to get:

$$y' - \frac{1}{3}y = \frac{t}{6}y^4$$

$$y^{-4}y' - \frac{1}{3}y^{-3} = \frac{t}{6}$$

$$-\frac{1}{3}v' - \frac{1}{3}v = \frac{t}{6}$$

$$2v' + 2v = -t$$

$$v' + v = -\frac{t}{2}$$

Let's solve this differential equation with integrating factors. We have that $p(t)=1$, so $\mu(t)=e^{\int p(t)dt}=e^{\int 1dt}=e^t$. We multiply both sides of the differential equation above by our integrating factor and we have that:

$$e^t v' + e^t v = -\frac{1}{2}te^t$$

$$\frac{d}{dt}(e^t v) = -\frac{1}{2}te^t$$

$$e^t v = \int -\frac{1}{2}te^t \, dt$$

We can now use integration by parts to evaluate the integral on the right hand side. Let $u=-\frac{1}{2}t$ and $dv=e^t \, dt$. Then $du=-\frac{1}{2}dt$ and $v=e^t$. Therefore:

$$\int u \, dv = uv - \int v \, du$$

$$\int -te^t \, dt = -\frac{1}{2}te^t + \frac{1}{2}\int e^t \, dt$$

$$\int -te^t \, dt = -\frac{1}{2}te^t + \frac{1}{2}e^t$$

$$\int -te^t \, dt = \frac{1}{2}(1-t)e^t + C$$

Therefore we have that:

$$e^t v = \frac{1-t}{2}e^t + C$$

$$v = \frac{1-t}{2} + Ce^{-t}$$

Now we made the substitution that $v=y^{-3}$ earlier. Therefore:

$$\frac{1}{y^3} = \frac{1-t}{2} + Ce^{-t}$$

$$y^3 = \frac{1}{\dfrac{1-t}{2} + Ce^{-t}}$$

Example: Solve the differential equation $y' - 5y = -5ty^3$.

Solution: Once again, the differential equation above is indeed a Bernoulli equation. It's already in the appropriate form and $n = 3$, so let $v = y^{1-n} = y^{-2}$. Then $v' = -2y^{-3}y'$. We take the differential equation above and divide both sides by y^3 and apply our substitutions to get:

$$y^{-3}y' - 5y^{-2} = -5t$$

$$-\frac{1}{2}v' - 5v = -5t$$

$$v' + 10v = 10t$$

Let's solve this differential equation using integrating factors again. We have that $p(t) = 10$. Therefore $\mu(t) = e^{10t}$. Multiplying both sides of our differential equation above by this gives us:

$$e^{10t}v' + 10e^{10t}v = 10te^{10t}$$

$$\frac{d}{dt}(e^{10t}v) = 10te^{10t}$$

$$e^{10t}v = 10\int te^{10t}dt$$

Let $u = t$ and $dv = e^{10t}dt$. Then $du = dt$ and $v = \frac{1}{10}e^{10t}$ so:

$$e^{10t}v = 10\left(\frac{t}{10}e^{10t} - \frac{1}{10}\int e^{10t}dt\right)$$

$$e^{10t}v = te^{10t} - \frac{1}{10}e^{10t} + C$$

$$v = t - \frac{1}{10} + Ce^{-10t}$$

Now we have that $v = y^{-2}$ and so:

$$y^{-2} = t - \frac{1}{10} + Ce^{-10t}$$

$$y^2 = \frac{1}{t - \dfrac{1}{10} + Ce^{-10t}}$$

Example: Solve the following IVP and find the interval of validity for the solution.

$$y' + \frac{4}{x}y = x^3y^2 \qquad y(2) = -1, \qquad x > 0$$

So, the first thing that we need to do is get this into the "proper" form and that means dividing everything by y^2. Doing this gives,

$$y^{-2}y' + \frac{4}{x}y^{-1} = x^3$$

The substitution and derivative that we'll need here is,

$$v = y^{-1} \qquad v' = -y^{-2}y'$$

With this substitution the differential equation becomes,

$$-v' + \frac{4}{x}v = x^3$$

Here's the solution to this differential equation.

$$v' - \frac{4}{x}v = -x^3 \qquad \Rightarrow \qquad \mu(x) = e^{\int -\frac{4}{x}dx} = e^{-4\ln|x|} = x^{-4}$$

$$\int (x^{-4}v)'dx = \int -x^{-1}dx$$

$$x^{-4}v = -\ln|x| + c \Rightarrow v(x) = cx^4 - x^4 \ln x$$

We can drop the absolute value bars on the xx in the logarithm because of the assumption that $x > 0$.

Now we need to determine the constant of integration. This can be done in one of two ways. We can convert the solution above into a solution in terms of y and then use the original initial condition or we can convert the initial condition to an initial condition in terms of v and use that. Because we'll need to convert the solution to y's eventually anyway and it won't add that much work in we'll do it that way.

So, to get the solution in terms of y all we need to do is plug the substitution back in. Doing this gives,

$$y^{-1} = x^4(c - \ln x)$$

At this point we can solve for y and then apply the initial condition or apply the initial condition and then solve for y. We'll generally do this with the later approach so let's apply the initial condition to get,

$$(-1)^{-1} = c2^4 - 2^4 \ln 2 \qquad \Rightarrow \qquad c = \ln 2 - \frac{1}{16}$$

Plugging in for c and solving for y gives,

$$y(x) = \frac{1}{x^4(\ln 2 - \frac{1}{16} - \ln x)} = \frac{-16}{x^4(1 + 16\ln x - 16\ln 2)} = \frac{-16}{x^4(1 + 16\ln \frac{x}{2})}$$

A little simplification can be done in the solution which will help in finding the interval of validity.

Before finding the interval of validity however, we mentioned above that we could convert the original initial condition into an initial condition for v. Let's briefly talk about how to do that. To do that all we need to do is plug $x = 2$ into the substitution and then use the original initial condition. Doing this gives,

$$v(2) = y^{-1}(2) = (-1)^{-1} = -1$$

So, in this case we got the same value for v that we had for y. Don't expect that to happen in general if you chose to do the problems in this manner.

Okay, let's now find the interval of validity for the solution. First, we already know that $x > 0$ and that means we'll avoid the problems of having logarithms of negative numbers and division by zero at $x = 0$. So, all that we need to worry about then is division by zero in the second term and this will happen where,

$$1 + 16 \ln \frac{x}{2} = 0$$

$$\ln \frac{x}{2} = -\frac{1}{16}$$

$$\frac{x}{2} = e^{-\frac{1}{16}} \Rightarrow x = 2e^{-\frac{1}{16}} \approx 1.8788$$

The two possible intervals of validity are then,

$$0 < x < 2e^{-\frac{1}{16}} \qquad 2e^{-\frac{1}{16}} < x < \infty$$

and since the second one contains the initial condition we know that the interval of validity is then $2e^{-\frac{1}{16}} < x < \infty$.

Here is a graph of the solution.

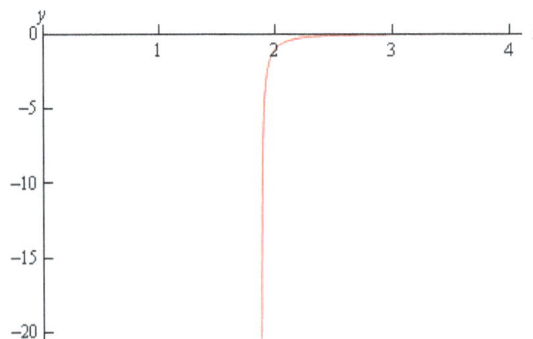

Example: Solve the following IVP and find the interval of validity for the solution.

$$y' = 5y + e^{-2x} y^{-2} \qquad y(0) = 2$$

The first thing we'll need to do here is multiply through by y^2 and we'll also do a little rearranging to get things into the form we'll need for the linear differential equation. This gives,

$$y^2 y' - 5y^3 = e^{-2x}$$

The substitution here and its derivative is

$$v = y^3 \qquad v' = 3y^2 y'$$

Plugging the substitution into the differential equation gives,

$$\frac{1}{3}v' - 5v = e^{-2x} \qquad \Rightarrow v' - 15v = 3e^{-2x} \qquad \mu(x) = e^{-15x}$$

We rearranged a little and gave the integrating factor for the linear differential equation solution. Upon solving we get,

$$v(x) = ce^{15x} - \frac{3}{17}e^{-2x}$$

Now go back to y's.

$$y^3 = ce^{15x} - \frac{3}{17}e^{-2x}$$

Applying the initial condition and solving for c gives,

$$8 = c - \frac{3}{17} \qquad \Rightarrow \qquad c = \frac{139}{17}$$

Plugging in c and solving for y gives,

$$y(x) = \left(\frac{139e^{15x} - 3e^{-2x}}{17} \right)^{\frac{1}{3}}$$

There are no problem values of x for this solution and so the interval of validity is all real numbers. Here's a graph of the solution.

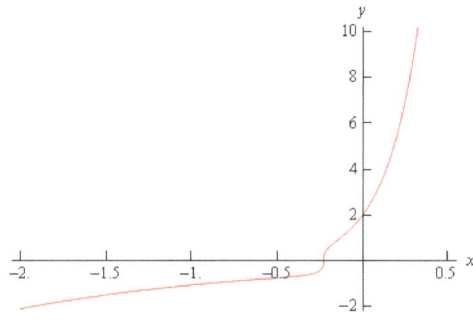

Example: Solve the following IVP and find the interval of validity for the solution.

$$6y' - 2y = xy^4 \qquad y(0) = -2$$

First get the differential equation in the proper form and then write down the substitution.

$$6y^{-4}y' - 2y^{-3} = x \qquad \Rightarrow \qquad v = y^{-3} \quad v' = -3y^{-4}y'$$

Plugging the substitution into the differential equation gives,

$$-2v' - 2v = x \qquad \Rightarrow \qquad v' + v = -\frac{1}{2}x \qquad \mu(x) = e^x$$

Again, we've rearranged a little and given the integrating factor needed to solve the linear differential equation. Upon solving the linear differential equation we have,

$$v(x) = -\frac{1}{2}(x-1) + ce^{-x}$$

Now back substitute to get back into y's.

$$y^{-3} = -\frac{1}{2}(x-1) + ce^{-x}$$

Now we need to apply the initial condition and solve for c.

$$-\frac{1}{8} = \frac{1}{2} + c \qquad \Rightarrow \qquad c = -\frac{5}{8}$$

Plugging in c and solving for y gives,

$$y(x) = -\frac{2}{(4x - 4 + 5e^{-x})^{\frac{1}{3}}}$$

Next, we need to think about the interval of validity. In this case all we need to worry about it is division by zero issues and using some form of computational aid (such as Maple or Mathematica) we will see that the denominator of our solution is never zero and so this solution will be valid for all real numbers.

Here is a graph of the solution.

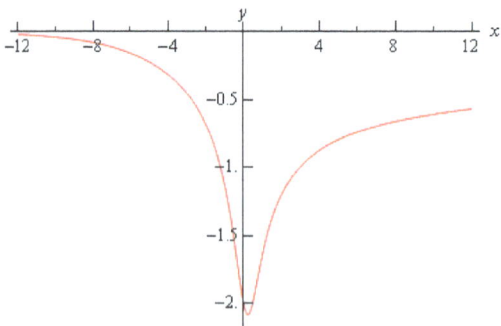

Example: Solve the following IVP and find the interval of validity for the solution.

$$y' + \frac{y}{x} - \sqrt{y} = 0 \qquad y(1) = 0$$

Let's first get the differential equation into proper form.

$$y' + \frac{1}{x}y = y^{\frac{1}{2}} \qquad \Rightarrow \qquad y^{-\frac{1}{2}}y' + \frac{1}{x}y^{-\frac{1}{2}} = 1$$

The substitution is then,

$$v = y^{\frac{1}{2}} \qquad v' = \frac{1}{2}y^{-\frac{1}{2}}y'$$

Now plug the substitution into the differential equation to get,

$$2v' + \frac{1}{x}v = 1 \qquad \Rightarrow \qquad v' + \frac{1}{2x}v = \frac{1}{2} \qquad \mu(x) = x^{\frac{1}{2}}$$

As we've done with the previous examples we've done some rearranging and given the integrating factor needed for solving the linear differential equation. Solving this gives us,

$$v(x) = \frac{1}{3}x + cx^{-\frac{1}{2}}$$

In terms of y this is,

$$y^{\frac{1}{2}} = \frac{1}{3}x + cx^{-\frac{1}{2}}$$

Applying the initial condition and solving for c gives,

$$0 = \frac{1}{3} + c \qquad \Rightarrow \qquad c = -\frac{1}{3}$$

Plugging in for c and solving for y gives us the solution.

$$y(x) = \left(\frac{1}{3}x - \frac{1}{3}x^{-\frac{1}{2}}\right)^2 = \frac{x^3 - 2x^{\frac{3}{2}} + 1}{9x}$$

We can multiply everything out and converted all the negative exponents to positive exponents to make the interval of validity clear here. Because of the root (in the second term in the numerator) and the x in the denominator we can see that we need to require $x > 0$ in order for the solution to exist and it will exist for all positive x's and so this is also the interval of validity.

Here is the graph of the solution.

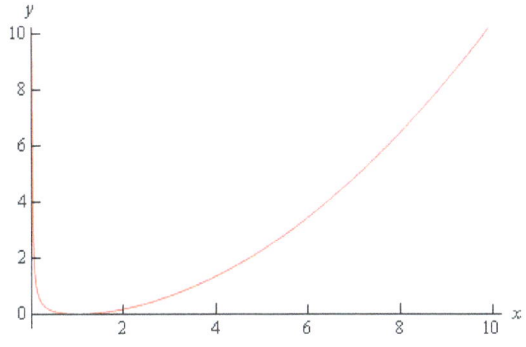

References

- Polyanin, Andrei D. (2001-11-28). Handbook of Linear Partial Differential Equations for Engineers and Scientists. Boca Raton, FL: Chapman & Hall/CRC. ISBN 1-58488-299-9.

- Ecuatii-diferentiale: alexnegrescu.files.wordpress.com, Retrieved 18 April 2018

- Homogeneous-Ordinary-Differential-Equation: mathworld.wolfram.com, Retrieved 28 June 2018

- Myint-U, Tyn; Debnath, Lokenath (2007). Linear Partial Differential Equations for Scientists and Engineers. Boston, MA: Birkhäuser Boston. doi:10.1007/978-0-8176-4560-1. ISBN 978-0-8176-4393-5. Retrieved 2011-03-29.

- Exact-differential-equation: math.tutorvista.com, Retrieved 16 May 2018

- First-order-equations/integrating-factors, differential-equations: cliffsnotes.com, Retrieved 26 March 2018

- Teschl, Gerald (2012). Ordinary Differential Equations and Dynamical Systems. Graduate Studies in Mathematics. 140. Providence, RI: American Mathematical Society. ISBN 978-0-8218-8328-0.

- The-integrating-factor-technique-5527: opencurriculum.org, Retrieved 16 May 2018

Partial Differential Equations

A differential equation, which contains unknown multivariable functions and their partial derivatives, is called a partial differential equation. All the diverse aspects of partial differential equations such as non-linear equations, Green's function, first- and second-order partial differential equation, fundamental solution, Euler–Tricomi equation, etc. have been carefully analyzed in this chapter.

A partial differential equation (or briefly a PDE) is a mathematical equation that involves two or more independent variables, an unknown function (dependent on those variables), and partial derivatives of the unknown function with respect to the independent variables. The *order* of a *partial differential equation* is the order of the highest derivative involved. A *solution* (or a *particular solution*) *to a partial differential equation* is a function that solves the equation or, in other words, turns it into an identity when substituted into the equation. A solution is called *general* if it contains all particular solutions of the equation concerned.

The term *exact solution* is often used for second- and higher-order nonlinear PDEs to denote a *particular solution*.

Partial differential equations are used to mathematically formulate, and thus aid the solution of, physical and other problems involving functions of several variables, such as the propagation of heat or sound, fluid flow, elasticity, electrostatics, electrodynamics, etc.

The symbol ∂ indicates a partial derivative, and is used when differentiating a function of two or more variables $u = u(x,t)$. For example means differentiate $u(x,t)$ with respect to t, treating x as a constant. Partial derivatives are as easy as ordinary derivatives.

There are three famous PDEs that you will encounter in our foundational course on PDEs. The second of the "big three" is the diffusion equations:

$$\frac{\partial u}{\partial t} - k \frac{\partial^2 u}{\partial x^2} = 0,$$

Where k is a constant. Here $u = u(x,t)$ is an unknown function of position and time.

If a drop of dye falls into a container of clear water it will gradually diffuse throughout the container. This process is described by the diffusion equation, with $u = u(x,t)$ representing the concentration of dye. This PDE also describes other processes of diffusion, for example the diffusion of heat. Think of holding a metal toasting fork in a camp fire. Eventually your hand will feel the heat. In this situation $u = u(x,t)$ represents the temperature of the fork.

The applications of the diffusion equation are not confined to science, however. Surprisingly, in recent years the diffusion equation has also played an important role in the area of mathematical finance.

The third member of the "big three" is Laplace's equation:

$$\frac{\partial^2 u}{\partial x^2} + \frac{\partial^2 u}{\partial y^2} = 0$$.

In addition to describing systems that evolve in time, PDEs also describe systems in a state of equilibrium, and here Laplace's equation comes into play.

As a simple example, think of a square sheet of metal, with three edges in contact with ice ($0°$ Celsius) and the fourth edge in contact with steam $(100°\ C)$. Heat will diffuse through the sheet, but eventually the temperature of the sheet will reach a steady state $u(x,y)$, that depends on position (x,y) but not on time. This function will satisfy Laplace's equation.

Basic Problems for PDEs of Mathematical Physics

Most PDEs of mathematical physics govern infinitely many qualitatively similar phenomena or processes. This follows from the fact that differential equations have, as a rule, infinitely many particular solutions. The specific solution that describes the physical phenomenon under study is separated from the set of particular solutions of the given differential equation by means of the initial and boundary conditions.

For simplicity and clarity of illustration, the basic problems of mathematical physics will be presented for the simplest linear equations $\dfrac{\partial w}{\partial t} - \dfrac{\partial^2 w}{\partial x^2} = 0$, $\dfrac{\partial^2 w}{\partial t^2} - \dfrac{\partial^2 w}{\partial x^2} = 0$, $\dfrac{\partial^2 w}{\partial x^2} + \dfrac{\partial^2 w}{\partial y^2} = 0$ only.

Cauchy problem and boundary value problems for parabolic equations

Cauchy problem $(t \geq 0, -\infty < x < \infty)$. Find a function w that satisfies heat equation $\dfrac{\partial w}{\partial t} - \dfrac{\partial^2 w}{\partial x^2} = 0$ for $t > 0$ and the initial condition

   $w = \varphi(x) \qquad at \qquad t = 0.$

The solution of the Cauchy problem is $\dfrac{\partial w}{\partial t} - \dfrac{\partial^2 w}{\partial x^2} = 0$, $w = \varphi(x) \qquad at \qquad t = 0.$

   $w(x,t) = \displaystyle\int_{-\infty}^{\infty} \varphi(\xi) E(x, \xi, t) d\xi,$

Where $E(x, \xi, t)$ is the *fundamental solution of the Cauchy problem,*

   $E(x, \xi, t) = \dfrac{1}{2\sqrt{\pi a t}} exp\left[-\dfrac{(x-\xi)^2}{4at}\right].$

In all *boundary value problems* (or *initial-boundary value problems*) below, it will be required to find a function w, in a domain $t \geq 0$, $x_1 \leq x \leq x_2$ $(-\infty < x_1 < x_2 < \infty)$, that satisfies the heat equation $\dfrac{\partial w}{\partial t} - \dfrac{\partial^2 w}{\partial x^2} = 0$ for $t > 0$ and the initial condition $w = \varphi(x) \qquad at \qquad t = 0.$ In addition, all problems will be supplemented with some boundary conditions as given below.

First boundary value problem. The function $w(x,t)$ takes prescribed values on the boundary:

$$w = \psi_1(t) \quad at \quad x = x_1,$$

$$w = \psi_2(t) \quad at \quad x = x_2$$

In particular, the solution to the first boundary value problem $\dfrac{\partial w}{\partial t} - \dfrac{\partial^2 w}{\partial x^2} = 0$ $w = \varphi(x)$ at $t = 0$,
$w = \psi_1(t) \quad at \quad x = x_1$
$w = \psi_2(t) \quad at \quad x = x_2$, with $\psi_1(t) = \psi_2(t) \equiv 0, x_1 = 0$, and $x_2 = l$ is expressed as

$$w(x,t) = \int_0^l \varphi(\xi) G(x, \xi, t) \, d\xi,$$

Where the Green's function $G(x, \xi, t)$ is defined by the formulas

$$G(x, \xi, t) = \frac{2}{l} \sum_{n=1}^{\infty} \sin\left(\frac{n\pi x}{l}\right) \sin\left(\frac{n\pi \xi}{l}\right) \exp\left(-\frac{an^2\pi^2 t}{l^2}\right)$$

$$= \frac{1}{2\sqrt{\pi a t}} \sum_{n=-\infty}^{\infty} \left\{ \exp\left[-\frac{(x-\xi+2nl)^2}{4at}\right] - \exp\left[-\frac{(x+\xi+2nl)^2}{4at}\right] \right\}.$$

The first series converges rapidly at large t and the second series at small t.

Second boundary value problem: The derivatives of the function $w(x,t)$ are prescribed on the boundary:

$$\frac{\partial w}{\partial x} = \psi_1(t) \qquad at \qquad x = x_1$$

$$\frac{\partial w}{\partial x} = \psi_2(t) \qquad at \qquad x = x_2.$$

Third boundary value problem: A linear relationship between the unknown function and its derivatives are prescribed on the boundary:

$$\frac{\partial w}{\partial x} - k_1 w = \psi_1(t) \qquad\qquad at \qquad x = x_1,$$

$$\frac{\partial w}{\partial x} + k_2 w = \psi_2(t) \qquad\qquad at \qquad x = x_2$$

Mixed boundary value problems: Conditions of different type, listed above, are set on the boundary of the domain in question, for example,

$$x = \psi_1(t) \qquad at \qquad x = x_1$$

$$\frac{\partial w}{\partial x} = \psi_2(t) \qquad at \qquad x = x_2$$

The boundary conditions given in above 4 equations are called homogeneous if $\psi_1(t) = \psi_2(t) \equiv 0$.

Solutions to the above initial-boundary value problems for the heat equation can be obtained by *separation of variables* (*Fourier method*) in the form of infinite series or by the *method of integral transforms* using the *Laplace transform*.

Cauchy problem and boundary value problems for hyperbolic equations:

Cauchy problem $(t \geq 0, -\infty < x < \infty)$. Find a function w that satisfies the wave equation $\dfrac{\partial^2 w}{\partial t^2} - \dfrac{\partial^2 w}{\partial x^2} = 0$ for $t > 0$ and two initial conditions

$$w = \varphi_0(x) \quad at \quad t = 0,$$

$$\frac{\partial w}{\partial t} = \varphi_1(x) \quad at \quad t = 0.$$

The solution of the Cauchy problem $\dfrac{\partial^2 w}{\partial t^2} - \dfrac{\partial^2 w}{\partial x^2} = 0$, $\begin{matrix} w = \varphi_0(x) & at & t = 0 \\ \dfrac{\partial w}{\partial t} = \varphi_1(x) & at & t = 0 \end{matrix}$, is given by *D'Alembert's formula*:

$$w(x,t) = \frac{1}{2}[\varphi_0(x + at) + \varphi_0(x - at)] + \frac{1}{2a}\int_{x-at}^{x+at} \varphi_1(\xi)d\xi.$$

Boundary value problems: In all *boundary value problems*, it is required to find a function w, in a domain $t \geq 0$, $x_1 \leq x \leq x_2$ $(-\infty < x_1 < x_2 < \infty)$, that satisfies the wave equation $\dfrac{\partial^2 w}{\partial t^2} - \dfrac{\partial^2 w}{\partial x^2} = 0$ for

0 and the initial conditions $\begin{matrix} w = \varphi_0(x) & at & t = 0 \\ \dfrac{\partial w}{\partial t} = \varphi_1(x) & at & t = 0 \end{matrix}$. In addition, appropriate boundary

conditions $\begin{matrix} w = \psi_1(t) & at & x = x_1 \\ w = \psi_2(t) & at & x = x_2 \end{matrix}$, $\begin{matrix} \dfrac{\partial w}{\partial x} = \psi_1(t) & at & x = x_1 \\ \dfrac{\partial w}{\partial x} = \psi_2(t) & at & x = x_2. \end{matrix}$, $\begin{matrix} \dfrac{\partial w}{\partial x} - k_1 w = \psi_1(t) & at & x = x_1 \\ \dfrac{\partial w}{\partial x} + k_2 w = \psi_2(t) & at & x = x_2 \end{matrix}$,

$x = \psi_1(t) \quad at \quad x = x_1$

$\dfrac{\partial w}{\partial x} = \psi_2(t) \quad at \quad x = x_2$ are imposed.

Solutions to these boundary value problems for the wave equation can be obtained by separation of variables (Fourier method) in the form of infinite series. In particular, the solution to the first boundary value problem $\dfrac{\partial^2 w}{\partial t^2} - \dfrac{\partial^2 w}{\partial x^2} = 0$, $\begin{matrix} w = \psi_1(t) & at & x = x_1, \\ w = \psi_2(t) & at & x = x_2 \end{matrix}$, $\begin{matrix} w = \varphi_0(x) & at & t = 0, \\ \dfrac{\partial w}{\partial x} = \varphi_1(x) & at & t = 0 \end{matrix}$

with homogeneous boundary conditions $\psi_1(t) = \psi_2(t) \equiv 0$ at $x_1 = 0$ and $x_2 = l$, is expressed as

$$w(x,t) = \frac{\partial}{\partial t}\int_0^l \varphi_0(\xi)G(x,\xi,t)\,d\xi + \int_0^l \varphi_1(\xi)G(x,\xi,t)\,d\xi,$$

where

$$G(x,\xi,t) = \frac{2}{a\pi}\sum_{n=1}^{\infty}\frac{1}{n}\sin\left(\frac{n\pi x}{l}\right)\sin\left(\frac{n\pi\xi}{l}\right)\sin\left(\frac{n\pi a t}{l}\right).$$

Goursat problem: On the characteristics of the wave equation $\dfrac{\partial^2 w}{\partial t^2} - \dfrac{\partial^2 w}{\partial x^2} = 0$, values of the unknown function w are prescribed:

$$w = \varphi(x) \quad \text{for} \quad x - t = 0 \quad (0 \le x \le a),$$

$$w = \psi(x) \quad \text{for} \quad x + t = 0 \quad (0 \le x \le b),$$

with the consistency condition $\varphi(0) = \psi(0)$ implied to hold.

Substituting the values set on the characteristics $\begin{matrix} w = \varphi(x+t) + \psi(x+t) & \text{for} & x - t = 0 & (0 \le x \le a), \\ w = \psi(x) & \text{for} & x + t = 0 & (0 \le x \le b), \end{matrix}$

into the general solution of the wave equation $w = \varphi(x+t) + \psi(x+t)$, one arrives at a system of linear algebraic equations for $\varphi(x)$ and $\psi(x)$. As a result, the solution to the Goursat problem $\dfrac{\partial^2 w}{\partial t^2} - \dfrac{\partial^2 w}{\partial x^2} = 0 , \begin{matrix} w = \varphi(x) & \text{for} & x - t = 0 & (0 \le x \le a), \\ w = \psi(x) & \text{for} & x + t = 0 & (0 \le x \le b) \end{matrix}$ is obtained in the form

$$w(x,t) = \varphi\left(\frac{x+t}{2}\right) + \psi\left(\frac{x-t}{2}\right) - \varphi(0).$$

The solution propagation domain is the parallelogram bounded by the four lines

$$x - t = 0, \qquad x + t = 0, \qquad x - t = 2b, \qquad x + t = 2a.$$

Boundary Value Problems for Elliptic Equations:

Setting boundary conditions for the first, second, and third boundary value problems for the Laplace equation $\dfrac{\partial^2 w}{\partial x^2} + \dfrac{\partial^2 w}{\partial y^2} = 0$ means prescribing values of the unknown function, its first derivative, and a linear combination of the unknown function and its derivative, respectively.

For example, the first boundary value problem in a rectangular domain $0 \le x \le a, 0 \le y \le b$ is characterized by the boundary conditions

$$w = \varphi_1(y) \quad at \quad x = 0, \qquad w = \varphi_2(y) \quad at \quad x = a,$$
$$w = \varphi_3(y) \quad at \quad y = 0, \qquad w = \varphi_4(y) \quad at \quad y = b.$$

The solution to problem $\dfrac{\partial^2 w}{\partial x^2} + \dfrac{\partial^2 w}{\partial y^2} = 0 , \begin{matrix} w = \varphi_1(y) & at & x = 0, & w = \varphi_2(y) & at & x = a, \\ w = \varphi_3(y) & at & y = 0, & w = \varphi_4(y) & at & y = b \end{matrix}$ with $\varphi_3(x) = \varphi_4(x) \equiv 0$ is given by

$$w(x,y) = \sum_{n=1}^{\infty} A_n \sinh\left[\frac{n\pi}{b}(a - x)\right] \sin\left(\frac{n\pi}{b}y\right) + \sum_{n=1}^{\infty} B_n \sinh\left(\frac{n\pi}{b}x\right) \sin\left(\frac{n\pi}{b}y\right),$$

Where the coefficients A_n and B_n are expressed as

$$A_n = \frac{2}{\lambda_n} \int_0^b \varphi_1(\xi) \sin\left(\frac{n\pi\xi}{b}\right) d\xi, \quad B_n = \frac{2}{\lambda_n} \int_0^b \varphi_2(\xi) \sin\left(\frac{n\pi\xi}{b}\right) d\xi, \quad \lambda_n = b \sinh\left(\frac{n\pi a}{b}\right)$$

Remark: For elliptic equations, the first boundary value problem is often called the *Dirichlet problem*, and the second boundary value problem is called the *Neumann problem*.

Some Nonlinear Equations Encountered in Applications

Nonlinear heat equation:

$$\frac{\partial w}{\partial t} = \frac{\partial}{\partial x}\left[f(w) \frac{\partial w}{\partial x} \right].$$

This equation describes one-dimensional unsteady thermal processes in quiescent media or solids in the case Where the thermal diffusivity is temperature dependent $f(w) > 0$. In the special case $f(w) \equiv 1$, the nonlinear equation $\frac{\partial w}{\partial t} = \frac{\partial}{\partial x}\left[f(w) \frac{\partial w}{\partial x} \right]$ becomes the linear heat equation $\frac{\partial w}{\partial t} - \frac{\partial^2 w}{\partial x^2} = 0$.

In general, the nonlinear heat equation $\frac{\partial w}{\partial t} = \frac{\partial}{\partial x}\left[f(w) \frac{\partial w}{\partial x} \right]$ admits exact solutions of the form

$$w = W(kx - \lambda t) \qquad (traveling - wave\ solution),$$

$$w = U(x / \sqrt{t}) \qquad (self - similar\ solution),$$

where $W = W(z)$ and $U = U(r)$ are determined by ordinary differential equations, and k and λ are arbitrary constants.

Kolmogorov–Petrovskii–Piskunov equation:

$$\frac{\partial w}{\partial t} = a \frac{\partial^2 w}{\partial x^2} + f(w), \qquad a > 0.$$

Equations of this form are often encountered in various problems of mass and heat transfer (with f being the rate of a volume chemical reaction), combustion theory, biology, and ecology.

In the special case of $f(w) \equiv 0$ and $a = 1$, the nonlinear equation $\frac{\partial w}{\partial t} = a \frac{\partial^2 w}{\partial x^2} + f(w), \qquad a > 0.$ becomes the linear heat equation $\frac{\partial w}{\partial t} - \frac{\partial^2 w}{\partial x^2} = 0$.

Remark: Equation $\frac{\partial w}{\partial t} = a \frac{\partial^2 w}{\partial x^2} + f(w), \qquad a > 0.$ is also called a *heat equation with a nonlinear source*.

Burgers equation:

$$\frac{\partial w}{\partial t} + w \frac{\partial w}{\partial x} = \frac{\partial^2 w}{\partial x^2}.$$

This equation is used for describing wave processes in gas dynamics, hydrodynamics, and acoustics.

1. Exact solutions to the Burgers equation can be obtained using the following formula (*Hopf–Cole transformation*):

$$w(x,t) = -\frac{2}{u}\frac{\partial u}{\partial x},$$

where $u = u(x,t)$ is a solution to the linear heat equation $u_t = u_{xx}$.

2. The solution to the Cauchy problem for the Burgers equation with the initial condition

$$w = f(x) \quad at \quad t = 0 \quad (-\infty < x < \infty)$$

has the form

$$w(x,t) = -2\,\frac{\partial}{\partial x}\ln F(x,t),$$

where

$$F(x,t) = \frac{1}{\sqrt{4\pi t}}\int_{-\infty}^{\infty}\exp\left[-\frac{(x-\xi)^2}{4t} + \frac{1}{2}\int_0^{\xi}f(\xi')d\xi'\right]d\xi.$$

Nonlinear wave equation:

$$\frac{\partial^2 w}{\partial t^2} = \frac{\partial}{\partial x}\left[f(w)\frac{\partial w}{\partial x}\right].$$

This equation is encountered in wave and gas dynamics $f(w) > 0$. In the special case $f(w) \equiv 1$, the nonlinear equation $\dfrac{\partial^2 w}{\partial t^2} = \dfrac{\partial}{\partial x}\left[f(w)\dfrac{\partial w}{\partial x}\right]$ becomes the linear wave equation $\dfrac{\partial^2 w}{\partial t^2} - \dfrac{\partial^2 w}{\partial x^2} = 0$.

Equation $\dfrac{\partial^2 w}{\partial t^2} = \dfrac{\partial}{\partial x}\left[f(w)\dfrac{\partial w}{\partial x}\right]$ admits exact solutions in implicit form:

$$x + t\sqrt{f(w)} = \varphi(w),$$
$$x - t\sqrt{f(w)} = \psi(w),$$

where $\varphi(w)$ and $\psi(w)$ are arbitrary functions.

Equation $\dfrac{\partial^2 w}{\partial t^2} = \dfrac{\partial}{\partial x}\left[f(w)\dfrac{\partial w}{\partial x}\right]$ can be reduced to a linear equation.

Nonlinear Klein–Gordon Equation

$$\frac{\partial^2 w}{\partial t^2} = a\frac{\partial^2 w}{\partial x^2} + f(w), \quad a > 0.$$

Equations of this form arise in differential geometry and various areas of physics (superconductivity, dislocations in crystals, waves in ferromagnetic materials, laser pulses in two-phase media, and others). For $f(w) \equiv 0$ and $a = 1$, equation $\dfrac{\partial^2 w}{\partial t^2} = a \dfrac{\partial^2 w}{\partial x^2} + f(w)$, $a > 0$ coincides with the linear wave equation $\dfrac{\partial^2 w}{\partial t^2} - \dfrac{\partial^2 w}{\partial x^2} = 0$.

1. In general, the nonlinear Klein–Gordon equation $\dfrac{\partial^2 w}{\partial t^2} = a \dfrac{\partial^2 w}{\partial x^2} + f(w)$, $a > 0$ admits exact solutions of the form

$$w = W(z), \quad z = kx - \lambda t,$$

$$w = U(\xi), \quad \xi = \left(\sqrt{at} + C_1\right)^2 - \left(x + C_2\right)^2,$$

Where $W = W(z)$ and $U = U(\xi)$ are determined by ordinary differential equations, while $k, \lambda, C_1,$ and C_2 are arbitrary constants.

2. In the special case

$$f(w) = be^{\beta w},$$

the general solution of equation $\dfrac{\partial^2 w}{\partial t^2} = a \dfrac{\partial^2 w}{\partial x^2} + f(w)$, $a > 0$ is expressed as

$$w(x,t) = \frac{1}{\beta}[\varphi(z) + \psi(y)] - \frac{2}{\beta} \ln \left| k \int \exp[\varphi(z)]dz - \frac{b\beta}{8ak} \int \exp[\psi(y)]dy \right|,$$

$$z = x - \sqrt{a}\, t, \quad y = x + \sqrt{a}\, t,$$

Where $\varphi = \varphi(z)$ and $\psi = \psi(y)$ are arbitrary functions and k is an arbitrary constant.

Remark: In   the   special   cases $f(w) = b\sin(\beta w)$ and $f(w) = b\sinh(\beta w)$,   equation $\dfrac{\partial^2 w}{\partial t^2} = a \dfrac{\partial^2 w}{\partial x^2} + f(w)$, $a > 0$ is called the sine-Gordon equation and the sinh-Gordon equation, respectively.

Nonlinear Laplace equation:

$$\frac{\partial^2 w}{\partial x^2} + \frac{\partial^2 w}{\partial y^2} = f(w).$$

This equation is also called a *stationary heat equation with a nonlinear source.*

1. In general, the nonlinear heat equation $\dfrac{\partial^2 w}{\partial x^2} + \dfrac{\partial^2 w}{\partial y^2} = f(w)$ admits exact solutions of the form

$$w = W(z), \quad z = k_1 x + k_2 y,$$

$$w = U(r), \quad r = \sqrt{(x + C_1)^2 + (y + C_2)^2},$$

Where $W = W(z)$ and $U = U(r)$ are determined by ordinary differential equations, while $k_1, k_2, C_1,$ and C_2 are arbitrary constants.

2. In the special case

$$f(w) = ae^{\beta w},$$

the general solution of equation $\dfrac{\partial^2 w}{\partial x^2} + \dfrac{\partial^2 w}{\partial y^2} = f(w).$ is expressed as

$$w(x, y) = -\frac{2}{\beta} \ln \frac{\left| 1 - 2a\beta \Phi(z)\overline{\Phi(z)} \right|}{4 \left| \Phi'_z(z) \right|}$$

Where $\Phi = \Phi(z)$ is an arbitrary analytic function of the complex variable $z = x + iy$ with nonzero derivative, and the bar over a symbol denotes the complex conjugate.

Monge–Ampere equation:

$$\left(\frac{\partial^2 w}{\partial x \partial y} \right)^2 - \frac{\partial^2 w}{\partial x^2} \frac{\partial^2 w}{\partial y^2} = f(x, y).$$

The equation is encountered in differential geometry, gas dynamics, and meteorology.

Below are solutions to the *homogeneous Monge–Ampere equation* for the special case $f(x, y) \equiv 0$.

1. Exact solutions involving one arbitrary function:

$$w(x, y) = \varphi(C_1 x + C_2 y) + C_3 x + C_4 y + C_5,$$

$$w(x, y) = (C_1 x + C_2 y) \, \varphi\left(\frac{y}{x} \right) + C_3 x + C_4 y + C_5,$$

$$w(x, y) = (C_1 x + C_2 y + C_3) \, \varphi\left(\frac{C_4 x + C_5 y + C_6}{C_1 x + C_2 y + C_3} \right) + C_7 x + C_8 y + C_9,$$

Where $C_1, \ldots, C_9$ are arbitrary constants and $\varphi = \varphi(z)$ is an arbitrary function.

2. General solution in parametric form:

$$w = tx + \varphi(t)y + \psi(t),$$

$$x + \varphi'(t)y + \psi'(t) = 0,$$

Where t is the parameter, and $\varphi = \varphi(t)$ and $\psi = \psi(t)$ are arbitrary functions.

Simplest Types of Exact Solutions of Nonlinear PDEs

Preliminary remarks

The following classes of solutions are usually regarded as exact solutions to nonlinear partial differential equations of mathematical physics:

- Solutions expressible in terms of elementary functions.

- Solutions expressed by quadrature.

- Solutions described by ordinary differential equations (or systems of ordinary differential equations).

- Solutions expressible in terms of solutions to linear partial differential equations (and/or solutions to linear integral equations).

The simplest types of exact solutions to nonlinear PDEs are traveling-wave solutions and self-similar solutions. They often occur in various applications.

In what follows, it is assumed that the unknown w depends on two variables, x and t, where t plays the role of time and x is a spatial coordinate.

Traveling-wave Solutions

Traveling-wave solutions, by definition, are of the form

$$w(x,t) = W(z), \qquad z = kx - \lambda t,$$

Where λ / k plays the role of the wave propagation velocity (the value $\lambda = 0$ corresponds to a stationary solution, and the value $k = 0$ corresponds to a space-homogeneous solution). Traveling-wave solutions are characterized by the fact that the profiles of these solutions at different time instants are obtained from one another by appropriate shifts (translations) along the x-axis. Consequently, a Cartesian coordinate system moving with a constant speed can be introduced in which the profile of the desired quantity is stationary. For $k > 0$ and $\lambda > 0$, the wave $w(x,t) = W(z)$, $z = kx - \lambda t$, travels along the x-axis to the right (in the direction of increasing x).

Traveling-wave solutions occur for equations that do not explicitly involve independent variables,

$$F\left(w, \frac{\partial w}{\partial x}, \frac{\partial w}{\partial t}, \frac{\partial^2 w}{\partial x^2}, \frac{\partial^2 w}{\partial x \, \partial t}, \frac{\partial^2 w}{\partial t^2}, \ldots\right) = 0.$$

Substituting $w(x,t) = W(z)$, $z = kx - \lambda t$, into $F\left(w, \dfrac{\partial w}{\partial x}, \dfrac{\partial w}{\partial t}, \dfrac{\partial^2 w}{\partial x^2}, \dfrac{\partial^2 w}{\partial x \, \partial t}, \dfrac{\partial^2 w}{\partial t^2}, \ldots\right) = 0.$, one obtains an autonomous ordinary differential equation for the function $W(z)$:

$$F(W, kW', -\lambda W', k^2 W'', -k\lambda W'', \lambda^2 W'', \ldots) = 0,$$

Where k and λ are arbitrary constants, and the prime denotes a derivative with respect to z.

Remark: The term *traveling-wave solution* is also used in the cases Where the variable t plays the role of a spatial coordinate, $t = y$.

All nonlinear equations considered above $\dfrac{\partial w}{\partial t} = \dfrac{\partial}{\partial x}\left[f(w)\dfrac{\partial w}{\partial x}\right]$, $\dfrac{\partial w}{\partial t} = a\dfrac{\partial^2 w}{\partial x^2} + f(w)$, $a > 0$,

$\dfrac{\partial w}{\partial t} + w\dfrac{\partial w}{\partial x} = \dfrac{\partial^2 w}{\partial x^2}$, $\dfrac{\partial^2 w}{\partial t^2} = \dfrac{\partial}{\partial x}\left[f(w)\dfrac{\partial w}{\partial x}\right]$, $\dfrac{\partial^2 w}{\partial t^2} = a\dfrac{\partial^2 w}{\partial x^2} + f(w)$, $a > 0$, $\dfrac{\partial^2 w}{\partial x^2} + \dfrac{\partial^2 w}{\partial y^2} = f(w)$ and

$w(x,t) = W(z)$, $z = kx - \lambda t$, with $f(x,y) = 0$, admit traveling-wave solutions.

Self-similar Solutions

By definition, a *self-similar solution* is a solution of the form

$$w(x,t) = t^{\alpha} U(\zeta), \qquad \zeta = x t^{\beta}.$$

The profiles of these solutions at different time instants are obtained from one another by a similarity transformation (like scaling).

Self-similar solutions exist if the scaling of the independent and dependent variables,

$$t = C\bar{t}, \quad x = C^k \bar{x}, w = C^m \bar{w}, \qquad \text{where } C \neq 0 \text{ is an arbitrary constant,}$$

for some k and m such that $|k| + |m| \neq 0$, is equivalent to the identical transformation.

It can be shown that the parameters in solution $w(x,t) = t^{\alpha} U(\zeta)$, $\zeta = x t^{\beta}$ and transformation $t = C\bar{t}$, $x = C^k \bar{x}, w = C^m \bar{w}$, where $C \neq 0$ is an arbitrary constant, are linked by the simple relations

$$\alpha = m, \qquad \beta = -k.$$

In practice, the above existence criterion is checked and if a pair of k and m in $t = C\bar{t}$, $x = C^k \bar{x}, w = C^m \bar{w}$, where $C \neq 0$ is an arbitrary constant, has been found, then a self-similar solution is defined by formulas $w(x,t) = t^{\alpha} U(\zeta)$, $\zeta = x t^{\beta}$ with parameters $\alpha = m$, $\beta = -k$.

Example. Consider the heat equation with a nonlinear power-law source term

$$\frac{\partial w}{\partial t} = a \frac{\partial^2 w}{\partial x^2} + b w^n.$$

The scaling transformation $t = C\bar{t}$, $x = C^k \bar{x}, w = C^m \bar{w}$, where $C \neq 0$ is an arbitrary constant, converts equation $\dfrac{\partial w}{\partial t} = a \dfrac{\partial^2 w}{\partial x^2} + b w^n$ into

$$C^{m-1} \frac{\partial \bar{w}}{\partial \bar{t}} = a C^{m-2k} \frac{\partial^2 \bar{w}}{\partial \bar{x}^2} + b C^{mn} \bar{w}^n$$

In order that equation $C^{m-1} \dfrac{\partial \bar{w}}{\partial t} = a C^{m-2k} \dfrac{\partial^2 \bar{w}}{\partial \bar{x}^2} + b C^{mn} \bar{w}^n$ coincides with $\dfrac{\partial w}{\partial t} = a \dfrac{\partial^2 w}{\partial x^2} + b w^n$ one must require that the powers of C are the same, which yields the following system of linear algebraic equations for the constants k and m:

$$m - 1 = m - 2k = mn$$

This system admits a unique solution

$$k = \frac{1}{2},$$

$m = \dfrac{1}{1-n}$. Using this solution together with relations $w(x,t) = t^{\alpha}U(\zeta)$, $\zeta = xt^{\beta}$ and

$\alpha = m$, $\beta = -k$, one obtains self-similar variables in the form

$$w = t^{1/(1-n)}U(\zeta),   \zeta = xt^{-1/2}.$$

Inserting these into $\dfrac{\partial w}{\partial t} = a\dfrac{\partial^2 w}{\partial x^2} + bw^n$, one arrives at the following ordinary differential equation for $U(\zeta)$

$$aU''_{\zeta\zeta} + \frac{1}{2}\zeta U'_{\zeta} + \frac{1}{n-1}U + bU^n = 0.$$

Cauchy Problem and Boundary Value Problems for Nonlinear Equations

The Cauchy problem and boundary value problems for nonlinear equations are stated in exactly the same way as for linear.

Examples. The Cauchy problem for a nonlinear heat equation is stated as follows: find a solution to equation $\dfrac{\partial w}{\partial t} = \dfrac{\partial}{\partial x}\left[f(w)\dfrac{\partial w}{\partial x}\right]$ subject to the initial condition $w = \varphi(x)$ at $t = 0$.

The first boundary value problem for a nonlinear wave equation as follows: find a solution to equation $\dfrac{\partial^2 w}{\partial x^2} + \dfrac{\partial^2 w}{\partial y^2} = f(w)$. subject to the initial conditions $w = \varphi(x)$ at $t = 0$. and the boundary conditions $\begin{array}{ll} w = \psi_1(t) & \text{at} \qquad x = x_1, \\ w = \psi_2(t) & \text{at} \qquad x = x_2 \end{array}$

Problems for nonlinear PDEs are normally solved using numerical methods.

Higher-Order Partial Differential Equations

Apart from second-order PDEs, higher-order equations also quite often arise in applications. Below are only a few important examples of such equations with some of their solutions.

Higher-Order Linear Partial Differential Equations

Equation of transverse vibration of elastic rod:

$$\frac{\partial^2 w}{\partial t^2} + a^2\frac{\partial^4 w}{\partial x^4} = 0.$$

The equation has the following particular solutions:

$$w(x,t) = [A\sin(\lambda x) + B\cos(\lambda x) + C\sinh(\lambda x) + D\cos(\lambda x)]\sin(\lambda^2 at),$$
$$w(x,t) = [A_1\sin(\lambda x) + B_1\cos(\lambda x) + C_1\sinh(\lambda x) + D_1\cos(\lambda x)]\cos(\lambda^2 at),$$

Where $A, B, C, D, A_1, B_1, C_1, D_1$, and λ are arbitrary constants.

Biharmonic equation:

$$\Delta\Delta w = 0,$$

Where $\Delta\Delta$ is the biharmonic operator,

$$\Delta\Delta \equiv \Delta^2 = \frac{\partial^4}{\partial x^4} + 2\frac{\partial^4}{\partial x^2 \partial y^2} + \frac{\partial^4}{\partial y^4}.$$

The biharmonic equation $\Delta\Delta w = 0,$ is encountered in plane problems of elasticity (w is the Airy stress function). It is also used to describe slow flows of viscous incompressible fluids (w is the stream function).

Various representations of the general solution to equation $\Delta\Delta w = 0,$ in terms of harmonic functions include

$$w(x,y) = xu_1(x,y) + u_2(x,y),$$
$$w(x,y) = yu_1(x,y) + u_2(x,y),$$
$$w(x,y) = (x^2 + y^2)u_1(x,y) + u_2(x,y),$$

Where u_1 and u_2 are arbitrary functions satisfying the Laplace equation $\Delta u_k = 0$ $(k = 1,2)$.

Complex form of representation of the general solution:

$$w(x,y) = Re[\bar{z}f(z) + g(z)],$$

Where $f(z)$ and $g(z)$ are arbitrary analytic functions of the complex variable $z = x + iy$; $\bar{z} = x - iy, i^2 = -1$. The symbol $Re[A]$ stands for the real part of a complex quantity A .

Higher-Order Nonlinear Partial Differential Equations

Korteweg–de Vries equation:

$$\frac{\partial w}{\partial t} + \frac{\partial^3 w}{\partial x^3} - 6w\frac{\partial w}{\partial x} = 0.$$

It is used in nonlinear mechanics and theoretical physics for describing one-dimensional nonlinear dispersive nondissipative waves. In particular, the mathematical modeling of moderate-amplitude shallow-water surface waves is based on this equation.

Equation of a steady laminar boundary layer on a flat plate:

$$\frac{\partial w}{\partial y}\frac{\partial^2 w}{\partial x \partial y} - \frac{\partial w}{\partial x}\frac{\partial^2 w}{\partial y^2} = a\frac{\partial^3 w}{\partial y^3}.$$

Where w is the stream function.

Boussinesq equation:

$$\frac{\partial^2 w}{\partial t^2} + \frac{\partial}{\partial x}\left(w\frac{\partial w}{\partial x}\right) + \frac{\partial^4 w}{\partial x^4} = 0.$$

This equation arises in several physical applications: propagation of long waves in shallow water, one-dimensional nonlinear lattice-waves, vibrations in a nonlinear string, and ion sound waves in a plasma.

Equation of motion of a viscous fluid:

$$\frac{\partial w}{\partial y}\frac{\partial}{\partial x}(\Delta w) - \frac{\partial w}{\partial x}\frac{\partial}{\partial y}(\Delta w) = a\Delta\Delta w, \qquad \Delta w = \frac{\partial^2 w}{\partial x^2} + \frac{\partial^2 w}{\partial y^2}.$$

This is a two-dimensional stationary equation of motion of a viscous incompressible fluid—it is obtained from the Navier–Stokes equations by the introduction of the stream function w.

Non-linear Partial Differential Equations

Linear Partial Differential Equations is the equations for which the dependent variable (and its derivatives) appear in terms with degree at most one. Anything else is called nonlinear.

An equation of the form

$$F\left(x, u,..., D^\alpha u\right) = 0,$$

Where $x = (x_1,..., x_n) \in R^n$, $u = (u_1,..., u_m) \in R^m$, $F = (F_1,..., F_k) \in R^k$, $\alpha = (\alpha_1,..., \alpha_n)$, is a multi-index of non-negative integers $\alpha_1,..., \alpha_n$, and $D^\alpha = D_1^{\alpha_1} ... D_n^{\alpha_n}$, Where $D_i = \partial/\partial x_i$, $i = 1,..., n$. In the case of complex-valued functions a non-linear partial differential equation is defined similarly. If $k > 1$ one speaks, as a rule, of a vectorial non-linear partial differential equation or of a system of non-linear partial differential equations. The order of $F(x, u,..., D^\alpha u) = 0$, is defined as the highest order of a derivative occurring in the equation.

One of the best known non-linear equations is the Monge–Ampère equation

$$\det\left|\frac{\partial^2 u}{\partial x_i \partial x_j}\right| + \sum_{i,j=1}^{n} A_{ij}(x, u, Du)\frac{\partial^2 u}{\partial x_i \partial x_j} + B(x, u, Du) = 0;$$

here and below $Du = (D_1 u,..., D_n u)$.

If $k = m$ and F is differentiable with respect to its variables corresponding to the derivatives of highest order, then the type of $F(x, u,..., D^\alpha u) = 0$, is defined as that of the principal linear part of F relative to these derivatives. One assigns, in general, to such a derivative (or to the resulting differential equation) a correspondingly defined weight. For example, in the non-linear heat equation

$$\frac{\partial u}{\partial x_1} = f\left(x_1, x_2, u, \frac{\partial u}{\partial x_2}, \frac{\partial^2 u}{\partial x_2^2}\right)$$

with $\partial f / \partial p_{22} > 0, p_{22} \Leftrightarrow \partial^2 u / \partial x_2^2$, the derivative $\partial f / \partial p_{22}$ has weight 2.

Since linearization $F(x, u, ..., D^\alpha u) = 0$, of with respect to the highest derivatives proceeds in a neighbourhood of a fixed solution, the type of $F(x, u, ..., D^\alpha u) = 0$, (in contrast to a linear equation even at a fixed point x) may depend on this fixed solution. For example, the equation

$$\frac{\partial^2 u}{\partial x_1^2} + \frac{\partial^2 u}{\partial x_2^2}\frac{\partial^2 u}{\partial x_2} = f(x_1, x_2)$$

is of elliptic type on solutions with $\partial u / \partial x_2 > 0$, but is of hyperbolic type on solutions with $\partial u / \partial x_2 < 0$.

The type of an equation determines whether boundary value (mixed) problems for this equations are well-posed and influences the method for studying them.

If the function F depends linearly on its highest derivatives, then $F(x, u, ..., D^\alpha u) = 0$, is called a quasi-linear equation. For example $\dfrac{\partial^2 u}{\partial x_1^2} + \dfrac{\partial^2 u}{\partial x_2^2}\dfrac{\partial u}{\partial x_2} = f(x_1, x_2)$, is quasi-linear. Otherwise the equation is called an essentially non-linear equation. For example, the Monge–Ampère equation

$$\det\left|\frac{\partial^2 u}{\partial x_i \partial x_j}\right| + \sum_{i, j=1}^{n} A_{ij}(x, u, Du)\frac{\partial^2 u}{\partial x_i \partial x_j} + B(x, u, Du) = 0 \text{; is essentially non-linear.}$$

If the coefficients of the highest derivatives of a quasi-linear equation do not depend on the solution (or its derivatives), then the equation is called a weakly non-linear equation. For example, the equation

$$\Delta u = f(x. u, Du)$$

is weakly non-linear.

The distinction between quasi-linear and weakly non-linear partial differential equations bears a conditional character and does not reflect an intrinsic property of the equations. Weakly non-linear equations may have stronger non-linearity properties than quasi-linear and even essentially non-linear equations. For example, there exist weakly non-linear equations of the form $\Delta u = f(x. u, Du)$ which have countably many distinct solutions for a given Dirichlet boundary condition in a bounded domain.

Equations of the form $F(x, u, ..., D^\alpha u) = 0$, can be considered in the whole space $\mathbf{R}^n$ or in some subdomain of it. In the first case the definition of the solution space includes conditions on the behaviour of the solutions at infinity. In the case of a domain one imposes on the boundary or on a part of it one or more boundary conditions. These boundary conditions may also involve non-linear operators.

A non-linear partial differential equation together with a boundary condition (or conditions) gives rise to a non-linear problem, which must be considered in an appropriate function space. The choice of this space of solutions is determined by the structure of both the non-linear differential operator F in the domain and that of the boundary operators. The choice of the function space for a non-linear problem is an essential feature in the investigation of the problem. For example, to the non-linear problem

$$\sum_{|\alpha|=m}(-1)^{|\alpha|}D^{\alpha}\left(\left|D^{\alpha}u\right|^{p-1}\operatorname{sgn}D^{\alpha}u\right)=f(x)\ \text{ with }\ p>1$$

in a bounded domain $\Omega\subset R^{n}$ and

$$D^{\beta}u=0,\qquad\left|\beta\right|\leq m-1,\qquad\text{ on the boundary }\partial\Omega$$

corresponds the Sobolev space $\overset{\circ}{W}{}_{p}^{m}(\Omega)$. This problem has, for any function f from the dual space

$W_{q}^{-m}(\Omega)=(\overset{\circ}{W}{}_{p}^{m}(\Omega))'$ $q^{-1}+p^{-1}=1$ a unique solution u in $\overset{\circ}{W}{}_{p}^{m}(\Omega)$. Here and below $\overset{\circ}{W}{}_{p}^{m}(\Omega)$ is the

closure in the Sobolev space $W^{m}(\Omega)$ of the set of all infinitely-differentiable functions in Ω of compact support.

In the study of problems for non-linear partial differential equations one treats questions such as: the existence of a solution and the number of solutions, or the non-existence, and collapse or bifurcation (branching) of a solution. Another problem is the asymptotic behaviour of a solution when the argument tends to the boundary, in particular, to infinity in the case of unbounded domains. The theory of such equations has two aspects: a local and a global one. The local theory is relatively completely developed for general non-linear problems of elliptic, parabolic or hyperbolic type. This theory is based on the implicit-function theorem in non-linear functional analysis and on the general theory of linear problems of corresponding type.

In the case of boundary value (mixed) problems for non-linear parabolic, or hyperbolic, equations this local theory makes it possible to establish the solvability of the problem on a sufficiently small time interval or on a fixed interval under the condition that the data of the problem deviate (in the appropriate metric) by a fairly small amount from the data for a known solution (as a rule, the zero solution) of a nearby problem.

The global theory of non-linear problems is less completely developed, and then only for individual classes of equations.

Non-linear First-order Partial Differential Equations

For a broad class of quasi-linear scalar first-order equations of the form

$$\frac{\partial u}{\partial t}+\sum_{i=1}^{n}\frac{\partial}{\partial x_{i}}\phi_{i}(t,x,u)+\psi(t,x,u)=0$$

The existence and uniqueness problem for the Cauchy problem with an initial condition for $t=0$ has been established for all $t>0$.

For a narrower class of equations of the form $\dfrac{\partial u}{\partial t}+\sum_{i=1}^{n}\dfrac{\partial}{\partial x_{i}}\phi_{i}(t,x,u)+\psi(t,x,u)=0$ the question

of the asymptotic behaviour of the solutions of such problems as $t\to+\infty$ and boundary value problems have also been discussed.

The theory of systems of quasi-linear first-order partial differential equations has been developed less completely.

Non-linear Second-order Partial Differential Equations

Equations of elliptic and parabolic type: The theory of global solvability of boundary value problems for a broad class of quasi-linear scalar second-order equations of elliptic type of the form

$$\sum_{i=1}^{n} \frac{\partial}{\partial x_i} \alpha_i (x, u, Du) + \alpha(x, u, Du) = 0,$$

or

$$\sum_{i,j=1}^{n} \alpha_{ij} (x, u, Du) \frac{\partial^2 u}{\partial x_i \partial x_j} + \alpha(x, u, Du) = 0$$

is relatively complete under the condition that an a priori estimate of $\max_x |u(x)|$ is available. Here the coefficients of the equations are subject to certain conditions.

A similar situation prevails in the theory of global solvability of boundary value (mixed) problems for the broad class of quasi-linear scalar second-order equations of parabolic type of the form

$$\frac{\partial u}{\partial t} = \sum_{i=1}^{n} \frac{\partial}{\partial x_i} \alpha_i (x, u, Du) + \alpha(x, u, Du),$$

or

$$\frac{\partial u}{\partial t} = \sum_{i,j} \alpha_{ij} (x, u, Du) \frac{\partial u}{\partial x_i \partial x_j} + \alpha(x, u, Du).$$

This solvability theory is based on a priori estimates and the Leray–Schauder method.

A priori estimates of $\max_x |u(x)|$ for certain classes of non-linear equations given in above 4 equation can be obtained on the basis of either the maximum principle or special integral inequalities and corresponding imbedding theorems for function spaces.

The global solvability theory of boundary value problems for essentially non-linear scalar equations of elliptic type has been developed for a narrow class of such equations in the case of two independent variables.

Global solvability of the boundary value (mixed) problem for essentially non-linear scalar equations of parabolic type has been established for the broad class of equations of the form

$$\frac{\partial u}{\partial t} = \alpha \left(x, u, \frac{\partial u}{\partial x}, \frac{\partial^2 u}{\partial x^2} \right)$$

in the case of a single space variable $x \in \mathbb{R}^n$.

Questions of global solvability of problems for systems of quasi-linear equations of elliptic or parabolic type have been considered only for narrow individual classes of such systems.

One of the effective methods in the study of second-order non-linear partial differential equations of either elliptic or parabolic type is the method of super- and subsolutions. Consider, for example, the boundary value problem

$$\begin{cases} -\Delta u = f\left(x, u, Du\right) & \text{in a bounded domain } \Omega \subset \mathbb{R}^n, \\ u = 0 & \text{on the boundary } \partial\Omega \text{ of class } C^2, \end{cases}$$

with a continuous function f defined on $\Omega \times \mathbb{R} \times \mathbb{R}^n$ and such that Bernshtein's inequality

$$\left|f\left(x, u, \xi\right)\right| \leq M\left(\left|u\right|\right) \cdot \left(1 + \left|\xi\right|^2\right)$$

holds for all $(x, u, \xi) \in \overline{\Omega} \times \mathbb{R} \times \mathbb{R}^n$ with an increasing function $M : \mathbb{R}_+ \to \mathbb{R}_+ = \{t \in \mathbb{R} : t \geq 0\}$. If there exist functions $u^+, u^- \in W_p^2(\Omega)$ with $p > n$ such that $-\Delta u^+ \geq f\left(x, u^+, Du^+\right)$ almost-everywhere in Ω, $u^+ \geq 0$ on $\partial\Omega$, $-\Delta u^- \leq f\left(x, u^-, Du^-\right)$ almost-everywhere in Ω, and $u^- \leq 0$ on $\partial\Omega$ (u^+ and u^- are super- and subsolutions of the problem $\begin{cases} -\Delta u = f\left(x, u, Du\right) & \text{in a bounded domain } \Omega \subset \mathbb{R}^n, \\ u = 0 & \text{on the boundary } \partial\Omega \text{ of class } C^2, \end{cases}$

and $u^+(x) \geq u^-(x)$ in Ω, then $\begin{cases} -\Delta u = f\left(x, u, Du\right) & \text{in a bounded domain } \Omega \subset \mathbb{R}^n, \\ u = 0 & \text{on the boundary } \partial\Omega \text{ of class } C^2, \end{cases}$ has a solution $u \in W_p^2(\Omega)$ and $u^-(x) \leq u(x) \leq u^+(x)$ in Ω. A good choice of super- and sub solutions of non-linear initial value problems based on solutions of model problems makes it possible to establish not only the solvability and a lower bound on the number of solutions, but also to obtain precise a priori estimates and the asymptotic behaviour of solutions of non-linear initial value problems.

The study of the global behaviour of solutions of boundary value (mixed) problems for non-linear parabolic equations is connected with that of stationary solutions of boundary value problems for corresponding non-linear elliptic equations, as it is in the case of ordinary differential equations.

Due to the fact that boundary value problems for non-linear elliptic equations do not always have a solution and that boundary value (mixed) problems for non-linear parabolic and hyperbolic equations need not have a solution for all $t > 0$, a theory of non-existence of solutions for non-linear partial differential equations has been developed.

Equations of hyperbolic type: These equations occupy a special place among the second-order non-linear differential equations. The "loss of one derivative" in the inversion of the second-order hyperbolic operator leads to principal obstacles in the study of non-linear hyperbolic equations. Even the local theory of non-linear hyperbolic equations and systems required the development of a special theory of implicit functions in non-linear functional analysis, since the classical implicit-function theorem in functional analysis turned out to be inapplicable here.

For (essentially) quasi-linear hyperbolic second-order equations in more than two independent variables the problem of global solvability, even for the Cauchy problem, has not been studied.

In the case of two independent variables ($t \in R_+$, $x \in R$) the global solvability of the Cauchy problem has been established for individual equations of the form

$$u_{tt} - \alpha^2(u_x)u_{xx} = f(t, x, u_t, u_x),$$

which reduce to special quasi-linear hyperbolic systems by the conservation principle.

In the case of quasi-linear hyperbolic equations of the form

$$u_{tt} - \alpha^2\left(\int_\Omega |Du|^2\, dx\right)\Delta u = f(t, x) \quad \text{in } \Omega \times [0, T],$$

the global solvability (for any $T > 0$) of the mixed problem with the conditions: $u = 0$ on $\partial\Omega \times [0, T]$ and $u = \phi$, $u_1 = \psi$ at $t = 0$, $x \in \Omega$, has been established in a certain class of smooth functions in x. Here Ω is a bounded domain in R^n with boundary $\partial\Omega$ of class C^∞ and $|Du|^2 = \sum_{i=1}^n (\partial u / \partial x_i)^2$.

The global solvability of the Cauchy problem and also of the boundary value (mixed) problem has been established for the broad class of weakly non-linear hyperbolic equations of the form

$$u_{tt} - \Delta u = f(t, x, u, u_t, Du), \qquad t > 0 \qquad x \in \Omega \subseteq R^n$$

A special place in the theory of non-linear hyperbolic equations is occupied by the problem of existence of periodic solutions (in t) of boundary value problems for these equations. Even for weakly non-linear equations this problem has been discussed only for equations of the form

$$u_{tt} - u_{xx} = f(t, x, u)$$

in the case of two variables $t \in R$ and $x \in [a, b] \subset R$. The complexity of this problem stems from the fact that the kernel of the corresponding linear problem is infinite dimensional.

Weakly non-linear and quasi-linear hyperbolic equations containing dissipative terms have been studied more completely.

Non-linear Partial Differential Equations of Higher Order

The solvability of boundary value (mixed) problems has been studied for the broad class of quasi-linear equations in divergence form

$$\sum_{|\alpha| \le m} (-1)^{|\alpha|} D^\alpha A_\alpha(x, u, \ldots, D^\beta u) = f(x),$$

$$\frac{\partial u}{\partial t} + \sum_{|\alpha| \le m} (-1)^{|\alpha|} D^\alpha A_\alpha(t, x, u, \ldots, D^\beta u) = f(t, x),$$

$$|\beta| \le m.$$

Concerning the functions A_α a number of conditions are assumed in this case which ensure that the non-linear operators are defined in the corresponding function spaces and satisfy certain conditions. For example, for the solvability of the boundary value problem in a bounded domain $\Omega \subset R^n$ with the conditions

$$D^w u = 0, \quad |w| \leq m-1, \quad \text{on the boundary } \partial\Omega,$$

$$\sum_{|\alpha| \leq m} (-1)^{|\alpha|} D^\alpha A_\alpha \left(x, u, ..., D^\beta u \right) = f(x),$$

for equation $\dfrac{\partial u}{\partial t} + \sum_{|\alpha| \leq m} (-1)^{|\alpha|} D^\alpha A_\alpha \left(t, x, u, ..., D^\beta u \right) = f(t, x)$, the following conditions are sufficient:

$$|\beta| \leq m.$$

1) The functions $A_\alpha \left(x, \xi_0, ..., \xi_\beta \right)$, $\xi_\beta \Leftrightarrow D^\beta u$, are measurable in x for all $\xi_0, ..., \xi_\beta$, continuous in $\left(\xi_0, ..., \xi_\beta \right)$ for almost-all $x \in \Omega$ and satisfy the inequality

$$\left| A_\alpha \left(x, \xi_0, ..., \xi_\beta \right) \right| \leq K \left[1 + \sum_{|\gamma| \leq m} |\xi_\gamma|^{p-1} \right]$$

with $p > 1$ and $K > 0$, $|\alpha| \leq n$.

2) The following coerciveness condition holds: For any function u in the Sobolev space $\overset{\circ}{W}{}_p^m(\Omega)$,

$$\sum_{|\alpha| \leq m} \int_\Omega A_\alpha \left(x, u, ..., D^\beta u \right) D^\alpha u \, dx \geq \alpha_0 \sum_{|\alpha| = m} \int_\Omega |D^\alpha u|^p \, dx - k \quad \text{with } \alpha_0 > 0.$$

3) The monotonicity condition holds: For any functions u and v in $\overset{\circ}{W}{}_p^m(\Omega)'$,

$$\sum_{|\alpha| \leq m} \int_\Omega A_\alpha \left(x, u, ..., D^\beta u \right) +$$

$$- A_\alpha \left(x, v, ..., D^\beta v \right))(D^\alpha u - D^\alpha v) \, dx \geq 0.$$

Under the above 3 conditions the boundary value problem $D^w u = 0$, $|w| \leq m-1$, on the boundary $\partial\Omega$

for equation

$$\sum_{|\alpha| \leq m} (-1)^{|\alpha|} D^\alpha A_\alpha \left(x, u, ..., D^\beta u \right) = f(x),$$

$\dfrac{\partial u}{\partial t} + \sum_{|\alpha| \leq m} (-1)^{|\alpha|} D^\alpha A_\alpha \left(t, x, u, ..., D^\beta u \right) = f(t, x)$, has a solution u in $\overset{\circ}{W}{}_p^m(\Omega)$ for any function f in the

$$|\beta| \leq m.$$

dual space $(\overset{\circ}{W}{}_p^m(\Omega))$.

All these conditions can be substantially relaxed. For example, the Fredholm alternative holds for

$$\sum_{|\alpha| \leq m} (-1)^{|\alpha|} D^\alpha A_\alpha \left(x, u, \ldots, D^\beta u \right) = f \left(x \right),$$

differential operators of the form $\dfrac{\partial u}{\partial t} + \sum_{|\alpha| \leq m} (-1)^{|\alpha|} D^\alpha A_\alpha \left(t, x, u, \ldots, D^\beta u \right) = f \left(t, x \right),$ with boundary

$$|\beta| \leq m.$$

conditions $D^w u = 0,$ $\quad |w| \leq m - 1,$ on the boundary $\partial \Omega$ that are odd and homogeneous, in princi-

ple, i.e. subject to certain conditions but without coercivity. If in this case the boundary value

problem with zero boundary conditions $D^w u = 0,$ $\quad |w| \leq m - 1,$ on the boundary $\partial \Omega$ for equation

$$\sum_{|\alpha| \leq m} (-1)^{|\alpha|} D^\alpha A_\alpha \left(x, u, \ldots, D^\beta u \right) = f \left(x \right),$$

$\dfrac{\partial u}{\partial t} + \sum_{|\alpha| \leq m} (-1)^{|\alpha|} D^\alpha A_\alpha \left(t, x, u, \ldots, D^\beta u \right) = f \left(t, x \right),$ with $f = 0$ has only the trivial solution, then this

$$|\beta| \leq m.$$

problem is solvable for any function f in the corresponding dual space.

For a broad class of boundary value (mixed) problems for non-linear partial differential equations a theory of normal solvability has been developed, generalizing the (Hausdorff) theory of normal solvability of linear operator equations to the non-linear case. This theory gives sufficient conditions for the solvability of boundary value (mixed) problems for non-linear equations and systems of equations of parabolic type and for weakly non-linear equations and systems of hyperbolic type.

In the theory of boundary value problems for non-linear equations of elliptic type, the question of the existence of eigen functions occupies a special place. The theory of eigen functions of boundary value problems for quasi-linear equations of elliptic type has been developed for a fairly broad class of problems. In particular, the abstract Lyusternik–Shnirel'man theory on the existence of a countable set of eigen functions has been transferred to a broad class of problems.

The theory of non-linear equations of elliptic type of higher order and in the theory of systems of non-linear equations of elliptic type in more than two independent variables is the question of regularity of solutions of these equations and systems.

In the case of a scalar quasi-linear uniformly-elliptic second-order equation with sufficiently smooth coefficients satisfying together with their first derivatives certain growth conditions, the solution has an interior regularity exceeding the smoothness of the right-hand side by two derivatives.

In the case of quasi-linear equations of elliptic type of order higher than two, and systems of quasi-linear equations of elliptic type of order two or more, in more than two independent variables, the corresponding smoothness of the solutions does not hold everywhere in the domain, but almost everywhere. Under additional conditions one succeeds in making precise the dimensions of the sets of zero (Hausdorff) measure on which the smoothness of solutions is, in general, violated. For a special class of quasi-linear elliptic systems with bounded non-linearities it has been established that the solutions are regular everywhere in the domain.

A theory has been established of boundary value problems for a broad class of quasi-linear equations of infinite order in divergence form.

Theory of Exact Solutions

Among the methods for constructing exact solutions one reckons: methods based on the application of group theory in the analysis of non-linear partial differential equations; methods based on Lie–Bäcklund transformation; and methods based on the inverse problem of scattering theory.

The inverse-scattering method has made it possible to study a number of physically important equations, such as the non-linear Korteweg–de Vries equation

$$u_t - 6uu_x + u_{xxx} = 0;$$

the non-linear sine-Gordon equation

$$u_{tt} - u_{xx} + \sin u = 0;$$

the non-linear Schrödinger equation

$$i\psi_t + \psi_{xx} + k|\psi|^2 \psi = 0,$$

and a number of others in one space variable $x \in R$. By means of this method one has been able to consider particular non-linear equations like the Korteweg–de Vries equation in two space variables.

Types of Solution

Complete Integral

A solution of a nonlinear partial differential equation (NPDE) of the form

$$f(x, y, z, p, q) = 0,$$

Where $p = \dfrac{\partial z}{\partial x}$ and $q = \dfrac{\partial z}{\partial y}$ involving two arbitrary constants is called a complete integral, and it is of the form

$$F(x, y, z, a, b) = 0.$$

Integral Surface Passing Through a Given Curve

Let us determine the solution of the equation $F(x,y,z,p,q)=0$ which passes through a given curve Γ having parametric equations $x = x(t)$, $y = y(t)$, $z = z(t)$, t being the parameter. For an integral surface through the curve, it is either

(a) A particular case of the complete integral

$$f(x,y, z,a,b) = 0$$

Where the constants a or b are given arbitrary values, or

(b) A particular case of the general integral corresponding to $f(x,y,z,a,b) = 0$, i.e. the envelope of a one-parameter subsystem $f(x,y,z,a,b) = 0$, or

(c) The envelope of a two-parameter system $f(x,y,z,a,b) = 0$.

It is obvious that the solution does not fall into category (a) or (c). So we consider the case (b) only.

Suppose that a surface S of the type (b) passes through a curve Γ. Then at its every point, the envelope S is touched by a member of the subsystem. In particular, at each point P of the curve Γ we may suppose that it is touched by a member S_p, say, of the subsystem, and since S_p touches S at P, it also touches S at the same point. Thus, S is the envelope of a one-parameter subsystem of $f(x,y,z,a,b) = 0$ each of whose members touches the curve Γ, provided that such a subsystem exists. To determine S, we have to consider the subsystem made up of those members of the family $f(x,y,z,a,b) = 0$ touching the curve Γ. The points of intersection of the surface $f(x,y,z,a,b) = 0$ and the curve Γ are determined in terms of the parameter t by the equation

$$f\{x(t),y(t),z(t),a,b\} = 0$$

and the equation that the curve Γ touches the surface $f(x,y,z,a,b) = 0$ is that the equation $f\{x(t),y(t),z(t),a,b\} = 0$ has two equal roots, i.e. the equation $f\{x(t),y(t),z(t),a,b\} = 0$ and the equation

$$\frac{\partial}{\partial t} f\{x(t),y(t),z(t),a,b\} = 0$$

have a common root, the condition for which is the t-eliminant by $f\{x(t),y(t),z(t),a,b\} = 0$ and $\frac{\partial}{\partial t} f\{x(t),y(t),z(t),a,b\} = 0$ in the form

$$\phi(a,b) = 0$$

This can be factorized into a set of alternative equations

$$b = \psi_1(a), b = \psi_2(a), \cdots$$

each of which defines a subsystem of one-parameter. The envelope of each of these one-parameter subsystems is a solution of the problem.

Example: Find a complete integral of the equation $p^2x + qy = z$ and hence derive the equation of an integral surface of which the line $y = 1, x + z = 0$ is a generator.

Solution: Let $F(x,y,z,p,q) = z - p^2x - qy = 0$. Then Charpit's equations are

$$\frac{dx}{-2px} = \frac{dy}{-q} = \frac{dz}{-2p^2x - qy} = \frac{dp}{p^2 + p} = \frac{dq}{0}$$

The last equation gives $q = constant = a$, say. The given equation then gives

$$p = \left(\frac{z - ay}{x}\right)^{1/2} \quad dz = pdx + qdy \Rightarrow \frac{d(z - ay)}{(z - ay)^{1/2}} = \frac{dx}{x^{1/2}}$$

which, on integration, leads to $(z - ay)^{1/2} = x^{1/2} + b^{1/2}$, Where b is constant. This solution can be written as

$$(x + ay - z + b)^2 = 4bx$$

which is the complete integral.

Now, the parametric equations of the given line are

$$x = t, \quad y = 1, \quad z = -t.$$

The intersection of $(x + ay - z + b)^2 = 4bx$ and $x = t$, $y = 1$, $z = -t$ is determined by $(2t + a + b)^2 = 4bt$, i.e. by the equation $4t^2 + 4at + (a + b)^2 = 0$ which has equal roots if a $a^2 = (a + b)^2$, i.e. if $b = -2a$. The appropriate one-parameter subsystem is

$$(x + ay - z - 2a)^2 = -8ax, \; i.e.(y-2)^2 a^2 + 2\{x(y + 2) - z(y + 2)\}a + (x - z)^2 = 0$$

and has for its envelope $\{x(y + 2) - z(y - 2)\} = (y - 2)^2 (x - z)^2$, i.e. $xy = z(y - 2)$.

The function z defined by this equation is the solution of the problem.

Problem of Finding one Complete Integral from the other

Let us suppose that

$$f(x, y, z, a, b) = 0$$

be a complete integral. We now show that there exists another relation

$$g(x, y, z, c, d) = 0$$

involving two arbitrary constant c and d which is also a complete integral. On the surface $g(x, y, z, c, d) = 0$ we choose a curve Γ whose equations contain the constants c, d as independent parameters and then find the envelope of the one-parameter subsystem of $x = t$, $y = 1$, $z = -t$ touching the curve Γ. Since this solution contains two arbitrary constants, it is a complete integral.

Example: Show that the partial differential equation $2xz + q^2 = x(xp + yq)$ has a complete integral $z + a^2x = axy + bx^2$ and deduce that $x(y + cx)^2 = 4(z - dx^2)$ is also a complete integral.

Solution: Let $F(x, y, z, p, q) = 2xz + q^2 - x(xp + yq) = 0$. Then Charpit's equations are

$$\frac{dx}{-x^2} = \frac{dy}{2q - xy} = \frac{dz}{-px^2 + 2q^2 - qxy} = \frac{dp}{-2z + yq} = \frac{dq}{-qx}.$$

From the first and last equations, we get on integration $q = ax$, where a is constant. Then the equation $F(x, y, z, p, q) = 0$ gives $p = (2z + a^2x - axy)/x$, so that the relation $dz = pdx + qdy$

leads to

$$dz = \frac{2z + a^2x - axy}{x}\, dx + axdy \;\Rightarrow\; d\left(\frac{z}{x^2}\right) + d\left(\frac{a^2}{x}\right) = a\, d\left(\frac{y}{x}\right)$$

which, on integration, gives $\dfrac{z}{x^2} + \dfrac{a^2}{x} = a\,\dfrac{y}{x} + b$, i.e. $z + a^2x = axy + bx^2$ which is a complete integral, b being a constant.

Now we show that

$$x(y + cx)^2 = 4(z - dx^2)$$

is also a complete integral. Let us consider the curve $\Gamma : y = 0, z = \dfrac{c^2x^3 + 4dx^2}{4}$ on the surface $x(y + cx)^2 = 4(z - dx^2)$. At the intersections of $z + a^2x = axy + bx^2$ and the curve Γ, we have

$$c^2x^2 + 4(d - b)x + 4a^2 = 0$$

Which has equal roots when $b = d \pm ac$. Taking $b = d + ac$, the subsystem has the equation $z + a^2x = axy + (d + ac)x^2$, i.e. $a^2x - x(cx + y)a + (z - ax^2) = 0$ which has the envelope $x^2(cx + y)^2 = 4x(z - dx^2)$, i.e. $x(y + cx)^2 = 4(z - dx^2)$. This is, therefore, a complete integral.

Integral Surface Circumscribing a given Surface

Two surfaces circumscribe each other if they touch along a curve, e.g. a conicoid and its enveloping cylinder and this curve may not be a plane curve. Suppose the partial differential equation (1) : $F(x, y, z, p, q) = 0$ has a complete integral of the form (2) : $f(x, y, z, a, b) = 0$. We determine, by the use of $f(x, y, z, a, b) = 0$, an integral surface of $F(x, y, z, p, q) = 0$ circumscribing the surface

$$\Sigma : \phi(x, y, z) = 0$$

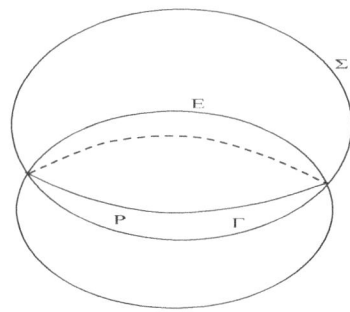

Consider the surface

$$E : \psi(x, y, z) = 0$$

of the required kind. Then it falls under the category (b) explained earlier. Since the possibility of occurrence of (b) is frequent, so we consider this type. Now E is the envelope of a one-parameter subsystem S of a two-parameter system $f(x,y,\ z,a,b)=0$ and, so it is touched at each point P of Γ by a member S_p of the subsystem S. Since S_p touches E at P, so it also touches $\sum$ at P. Hence the equation $E:\psi(x,y,\ z)\ =\ 0$ represents the envelope of the set of surfaces $f(x,y,\ z,a,b)=0$ which touch the surface $\sum:\phi(x,y,\ z)\ =\ 0$.

Let us now find the surface $f(x,y,\ z,a,b)=0$ which touches E and see whether this provides a solution of the equation $F(x,y,\ z,\ p,q)\ =\ 0$. The surface $f(x,y,\ z,a,b)=0$ touches the surface $\sum:\phi(x,y,\ z)\ =\ 0$ if the equations $f(x,y,\ z,a,b)=0$,

$\sum:\phi(x,y,\ z)\ =\ 0$ and

$$\frac{f_x}{\phi_x}=\frac{f_y}{\phi_y}=\frac{f_z}{\phi_z}$$

are consistent, the condition for which is that the elimination of $x,\ y,\ z$ from the four equations yielding a relation of the form

$$\chi(a,b)=0$$

between a and b. This equation can be factorised into a set of relations of the type

$$b=X_1(a),\ b=X_2(a),...$$

which defines a subsystem of $f(x,y,\ z,a,b)=0$ each of which touches $\sum:\phi(x,y,\ z)\ =\ 0$. The points of contact lie on the surface whose equation is the $a,\ b$ -eliminant of the equations $\frac{f_x}{\phi_x}=\frac{f_y}{\phi_y}=\frac{f_z}{\phi_z}$ and $b=X_1(a),\ b=X_2(a),....$ The intersection of this surface with $\sum$ gives the curve Γ and each of the relations $b=X_1(a),\ b=X_2(a),...$ defines a subsystem whose envelope E touches $\sum$ along Γ.

Example: Show that the integral surface of the equation $2y(1+p^2)=pq$ whose equation is circumscribed about the cone $x^2+z^2=y^2$ has the equation $z^2=y^2(4y^2+4x+1)$.

Solution: Let $F(x,y,\ z,\ p,q)\ =\ 2y(1+p^2\ )-pq\ =0$. Then Charpit's equations are

$$\frac{dx}{4yp-q}=\frac{dy}{-p}=\frac{dz}{4yp^2-2pq}=\frac{dp}{0}=\frac{dq}{-(2\ +\ 2p^2)}$$

The fourth equation gives $p\ =\ $ constant $\ =\ a$, say, and then from the given equation, we get $q=\dfrac{2y(1+a^2)}{a}$. Thus the equation $dz=pdx+qdy$ leads to

$$dz=adx+\frac{2y(1+a^2)}{a}\ dy\ \Rightarrow z\ =\ ax\ +\ \frac{y^2\ (1+a^2\ )}{a}+b$$

on integration, b being a constant. This is a complete integral.

Now, let $f(x, y, z, a, b) = z - ax - \dfrac{y^2(1+a^2)}{a} - b = 0$ and $\psi(x, y, z) = x^2 + y^2 - z^2 = 0$. Then from the relation

$$\frac{f_x}{\psi_x} = \frac{f_y}{\psi_y} = \frac{f_z}{\psi_z}, \quad \text{we get,} \quad \frac{-a}{2x} = \frac{-2y(1+a^2)}{-2ay} = \frac{1}{2z} \quad \text{so that}$$

$$x = -\frac{a^2}{2(1+a^2)}, z = \frac{a}{2(1+a^2)} \quad \text{and the equation } f = 0 \text{ gives } y^2 = \frac{a(a-2b)}{2(1+a^2)}.$$

Using these values of x, y, z, we find from the equation $\psi = 0$ that the relation $b = a/4$ defines a subsystem whose envelope is a surface of the required kind. The envelope of the subsystem is obviously $z^2 = y^2(4y^2 + 4x + 1)$.

General Integral

Any relation of the form

$$F(u, v) = 0$$

Involving an arbitrary function F connecting two known functions $u(x, y, z)$ and $v(x, y, z)$ and providing a solution of a first-order PDE is called a general solution or a general integral of that first-order PDE.

It is possible to derive a general integral of the PDE once a complete integral is known.

With $b = \phi(a)$, if we take any one-parameter subsystem

$$f(x, y, z, a, \phi(a)) = 0$$

of the system $F(x, y, z, a, b) = 0$ and form its envelope, we obtain a solution of equation $f(x, y, z, p, q) = 0$. When $\phi(a)$ is arbitrary, the solution obtained is called the general integral of $f(x, y, z, p, q) = 0$ corresponding to the complete integral $F(x, y, z, a, b) = 0$.

Singular Solution

A function $\varphi(x)$ is called the *singular solution* of the differential equation $F(x, y, y') = 0$ if *uniqueness of solution* is violated at each point of the domain of the equation. Geometrically this means that more than one integral curve with the common tangent line passes through each point (x_0, y_0).

Note: Sometimes the weaker definition of the singular solution is used, when the uniqueness of solution of differential equation may be violated only at some points.

A singular solution of a differential equation is not described by the general integral, that is it can not be derived from the general solution for any particular value of the constant C.

Example: Suppose that the following equation is required to be solved: $(y')^2 - 4y = 0$.

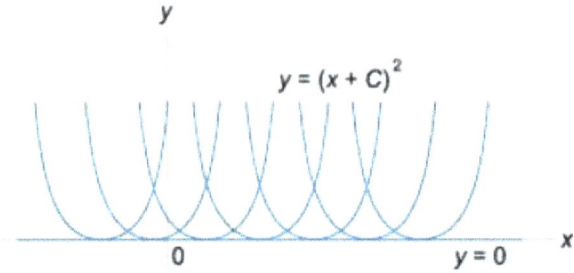

It is easy to see that the general solution of the equation is given by the function $y = (x + C)^2$. Graphically, it is represented by the family of parabolas in above figure.

Besides this, the function $y = 0$ also satisfies the differential equation. However, this function is not contained in the general solution! Since more than one integral curve passes through each point of the straight line $y = 0$, the uniqueness of solution is violated on this line, and hence it is a singular solution of the differential equation.

p-discriminant

One of the ways to find a singular solution is investigation of the so-called p-*discriminant* of the differential equation. If the function $F(x, y, y')$ and its partial derivatives $\dfrac{\partial F}{\partial y}, \dfrac{\partial F}{\partial y'}$ are continuous in the domain of the differential equation, the singular solution can be found from the system of equations:

$$\begin{cases} F(x, y, y') = 0 \\ \dfrac{\partial F(x, y, y')}{\partial y'} = 0 \end{cases}.$$

The equation $\psi(x, y) = 0$ obtained by solving the given system of equations is called the p-*discriminant* of the differential equation. The corresponding curve determined by this equation is called a p-*discriminant curve*.

Upon finding the p-discriminant curve, one should check the following:

- Whether it is a solution of the differential equation?

- Whether it is a singular solution, i.e. are there any other integral curves of the differential equation that touch the p-discriminant curve at each point?

This can be done as follows:

- Find the general solution of the differential equation (denote it by y_1);

- Write the conditions of touching the singular solution (denote it by y_2) and the general solution y_1 at an arbitrary point x_0:

$$\begin{cases} y_1(x_0) = y_2(x_0) \\ y_1'(x_0) = y_2'(x_0) \end{cases};$$

If the system has a solution at the arbitrary point x_0, the function y_2 is a singular solution. The singular solution usually corresponds to the *envelope* of the family of integral curves of the general solution of the differential equation.

Envelope of the Family of Integral Curves and C-discriminant

Another way to find a singular solution as the *envelope* of the family of integral curves is based on using C-*discriminant*.

Let $\Phi(x,y,C)$ be the general solution of a differential equation $F(x,y,y')=0$. Graphically the equation $\Phi(x,y,C)=0$ corresponds to the family of integral curves in the xy-plane. If the function $\Phi(x,y,C)$ and its partial derivatives are continuous, the envelope of the family of integral curves of the general solution is defined by the system of equations:

$$\begin{cases} \Phi(x,y,C) = 0 \\ \dfrac{\partial \Phi(x,y,C)}{\partial C} = 0 \end{cases}.$$

General Algorithm of Finding Singular Points

A more common way of finding singular points of a differential equation is based on the *simultaneous using* p-*discriminant and* C-*discriminant*.

Here first we find the equations of the p-discriminant and C-discriminant:

- $\psi_p(x,y)=0$ is the equation of the p-discriminant;

- $\psi_C(x,y)=0$ is the equation of the C-discriminant.

It turns out that these equations have a certain structure. In general case, the equation of the p-discriminant can be factored into the product of three functions:

$$\psi_P(x,y) = E \times T^2 \times C = 0,$$

WhereE means the equation of the *Envelope*, T is the equation of the *Tac locus*, and C is the equation of the *Cusp locus*.

Similarly, the equation of the C-discriminant can be also factored into the product of three functions:

$$\psi_C(x,y) = E \times N^2 \times C^3 = 0,$$

Where E is the equation of the *Envelope*, N is the equation of the *Node locus*, and C is the equation of the *Cusp locus*.

Here we meet with new kinds of singular points: C – *Cusp loci*, T- *Tac loci*, and N – *Node loci*. Their view in the xy-plane is shown schematically in Figures given below.

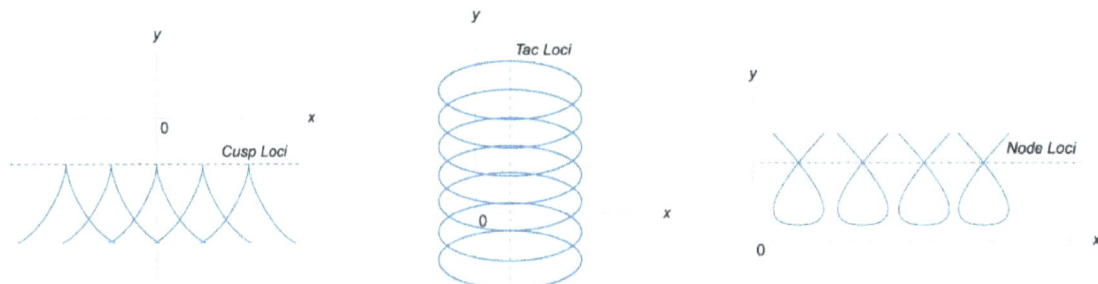

Three of the four types of points, namely, the Tac loci, Cusp loci and Node loci are extraneous points, i.e. they do not satisfy the differential equation and, therefore, they are not singular solutions of the differential equation. Only the envelope of the considered points is the singular solution. Since the envelope is presented in the equations of the both discriminants as a first degree factor, this allows to find the equation of the envelope.

Example:

Find the singular solutions of the equation $1+(y')^2 = \dfrac{1}{y^2}$.

Solution:

We will use p-*discriminant* for investigation of the singular points. Differentiating the equation with respect to y' gives:

$$2y' = 0, \Rightarrow y' = 0.$$

Putting this into the differential equation yields the equation of the p-discriminant:

$$1+0 = \frac{1}{y^2}.$$

It follows from here that the equation of the p-discriminant describes two horizontal lines: $y = \pm 1$ .It is easy to check that this solution satisfies the given differential equation:

$$y = \pm 1, \Rightarrow y' = 0, \Rightarrow 1+0^2 = \frac{1}{1^2}, \Rightarrow 1 = 1.$$

Now we find the general solution of the differential equation. We can write it in the following form:

$$(y')^2 = \frac{1}{y^2} - 1 = \frac{1-y^2}{y^2}, \Rightarrow y' = \pm\frac{\sqrt{1-y^2}}{y}, \Rightarrow \frac{ydy}{\sqrt{1-y^2}} = \pm dx.$$

Make the replacement:

$$1-y^2 = t, \Rightarrow -2ydy = dt, \Rightarrow ydy = -\frac{dt}{2}.$$

As a result, we get:

$$\frac{\left(-\dfrac{dt}{2}\right)}{\sqrt{t}} = \pm dx.$$

After integration we obtain the general solution of the differential equation:

$$\int \frac{dt}{2\sqrt{t}} = \pm \int dx, \quad \Rightarrow \sqrt{t} = \pm x + C, \quad \Rightarrow \sqrt{1-y^2} = \pm(x+C),$$

where C is an arbitrary constant.

The last expression can be written as follows:

$$(x+C)^2 + y^2 = 1.$$

This equation describes the family of circles of radius 1, filling in the band $-1 \le y \le 1$ below Figure. As it can be seen from the Figure, the p-discriminant lines $y = \pm 1$ are the envelopes for the given circles. However, we must prove formally that the uniqueness of solution is violated on these straight lines.

Take an arbitrary point x_0. Write the condition of touching two integral curves at this point:

$$\begin{cases} y_1(x_0) = y_2(x_0) \\ y_1'(x_0) = y_2'(x_0) \end{cases}.$$

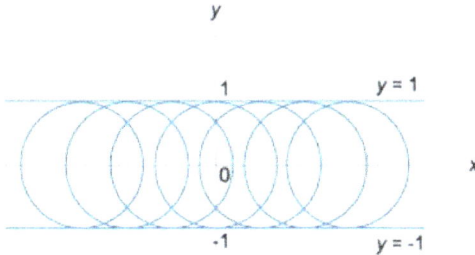

Here $y_1(x)$ denotes our general solution that has the form (for the upper semicircle):

$$y_1(x) = \sqrt{1-(x+C)^2}.$$

The function $y_2(x)$ corresponds to the horizontal line $y = 1$. Both the lines will touch at the point x_0 if only the following relations are satisfied:

$$\begin{cases} \sqrt{1-(x_0+C)^2} = 1 \\ \dfrac{-x_0-C}{\sqrt{1-x_0+C}} = 0 \end{cases}.$$

The given conditions are satisfied, if we set $C = -x_0$.

Thus, we have proved that at each point xo on the straight line $y = 1$ there exists a touching circle with $C = -x_0$. Hence, the uniqueness of solution is violated at each point of the straight line. Therefore, the line $y = 1$ is a singular solution of the given differential equation. Similarly, we can prove that the line $y = -1$ is also a singular solution.

Example:

Find the singular solution of the differential equation $y = (y')^2 - 3xy' + 3x^2$. The general solution of the equation is known and given by the function $y = Cx + C^2 + x^2$.

Solution:

We will use C-*discriminant* to determine the singular solution. Since the general solution of the differential equation is known, we can write:

$$\Phi(x, y, C) = y - Cx - C^2 - x^2.$$

The partial derivative in C is

$$\frac{\partial \Phi(x, y, C)}{\partial C} = -x - 2C.$$

We get the following system of equations:

$$\begin{cases} y = Cx - C^2 - x^2 \\ -x - 2C = 0 \end{cases}.$$

It follows from the second equation that $C = -\dfrac{x}{2}$. Substituting this in the first equation, we find the C-discriminant curve, which is a parabola:

$$y = \left(-\frac{x}{2}\right) \cdot x - \left(-\frac{x}{2}\right)^2 - x^2 = \frac{3}{4}x^2.$$

Make sure that this function is a solution of the original differential equation:

$$y = \frac{3}{4}x^2, \ \Rightarrow y' = \frac{3}{2}x, \ \Rightarrow \frac{3}{4}x^2 = \left(\frac{3}{2}x\right)^2 - 3x \cdot \frac{3}{2}x + 3x^2,$$

$$\Rightarrow \frac{3}{4}x^2 = \frac{9}{4}x^2 - \frac{9}{2}x^2 + 3x^2, \ \Rightarrow \frac{3}{4}x^2 = \frac{3}{4}x^2$$

Now we check that uniqueness of solution is violated on this curve. Denote:

$$y_1 = Cx + C^2 + x^2, y_2 = \frac{3}{4}x^2.$$

Write the conditions of touching the two curves at a certain arbitrary point x_0 as

$$\begin{cases} y_1(x_0) = y_2(x_0) \\ y_1'(x_0) = y_2'(x_0) \end{cases}.$$

As a result, we have:

$$\begin{cases} Cx_0 + C^2 + x_0^2 = \dfrac{3}{4}x_0^2 \\ C + 2x_0 = \dfrac{3}{2}x_0 \end{cases}.$$

The given system of equations is consistent when the constant C at each point x_0 is equal to

$$C = -\frac{x_0}{2}.$$

Thus, we have proved that the C-discriminant curve $y = \dfrac{3}{4}x^2$ is the envelope (i.e. the singular solution) for the family of parabolas $y = Cx + C^2 + x^2$ representing the general solution of the differential equation.

Example:

Investigate the singular solutions of the differential equation $(y')^2(1-y)^2 = 2-y$.

Solution:

First we find p-*discriminant* of the given equation. Differentiating with respect to x gives:

$$2y'(1-y)^2 = 0.$$

Eliminate y' from the system of the equations:

$$\begin{cases} (y')^2(1-y)^2 = 2-y \\ y'(1-y)^2 = 0 \end{cases}.$$

Then we obtain:

$$(y')^2 = \frac{2-y}{(1-y)^2}, \quad \Rightarrow \frac{2-y}{(1-y)^2}\cdot(1-y)^4 = 0,$$
$$\Rightarrow (1-y)^2(2-y) = 0$$

Now we determine C-*discriminant*. Unfortunately, to find it, we need to solve the differential equation and find its general solution . Rewrite the equation in the following form:

$$\left(\frac{dy}{dx}\right)^2 = \frac{2-y}{(1-y)^2}, \quad \Rightarrow \frac{dy}{dx} = \pm\frac{\sqrt{2-y}}{1-y}, \quad \Rightarrow \frac{(1-y)dy}{\sqrt{2-y}} = \pm dx$$

By integrating both sides we have:

$$\int \frac{(1-y)dy}{\sqrt{2-y}} = \pm \int dx + C.$$

We change the variable in the left integral:

$$2 - y = t, \quad \Rightarrow dy = -dt, \quad \Rightarrow 1 - y = t - 1.$$

This yields:

$$\int \frac{(t-1)(-dt)}{\sqrt{t}} = \pm x + C, \Rightarrow \int \left(\sqrt{t} - \frac{1}{\sqrt{t}}\right) dt = \mp x - C$$

$$\Rightarrow \frac{t^{\frac{3}{2}}}{\frac{3}{2}} - \frac{t^{\frac{1}{2}}}{\frac{1}{2}} = \mp x - C, \Rightarrow \frac{2}{3}t^{\frac{3}{2}} - 2t^{\frac{1}{2}} = \mp x - C,$$

$$\Rightarrow \frac{2}{3}(2-y)^{\frac{3}{2}} - 2(2-y)^{\frac{1}{2}} = \mp x - C,$$

$$\Rightarrow \frac{2}{3}\sqrt{2-y}(2-y-3) = \mp x - C,$$

$$\Rightarrow \frac{4}{9}(2-y)(y+1)^2 = (x+C)^2,$$

$$\Rightarrow 4(2-y)(y+1)^2 = 9(x+C)^2$$

Differentiate the general solution with respect to C:

$$0 = 18(x+C).$$

Putting $x + C = 0$ back into the general solution, we obtain the equation of the C-discriminant:

$$(y+1)^2 (2-y) = 0.$$

Now we can write both discriminants together:

$$\psi_p(y) = (1-y)^2(2-y) = 0,$$

$$\psi_C(y) = (y+1)^2(2-y) = 0.$$

As it follows from the structure of the discriminants, the equation $2 - y = 0$ is the equation of the *envelope* because it's contained in the both discriminants as the first degree factor. We can also find the equation of the *Tac locus* from the expression for p-discriminant:

$$(1-y)^2 = 0, \quad \Rightarrow y = 1.$$

And similarly, as it follows from the expression for C-discriminant, the equation the *Node locus* is given by

$$(y+1)^2 = 0, \Rightarrow y = -1.$$

In the given example, only the *envelope* $y = 2$ is the singular solution of the differential equation.

Charpit's Method

Charpit's method is a general method for finding the complete solution of non-linear partial differential equation of the first order of the form

$$f(x, y, z, p, q) = 0.$$

Since we know that $dz = \dfrac{\partial z}{\partial x}dx + \dfrac{\partial z}{\partial y}dy = pdx + qdy$.

Integrating $dz = \dfrac{\partial z}{\partial x}dx + \dfrac{\partial z}{\partial y}dy = pdx + qdy$, we get the complete solution of $f(x, y, z, p, q) = 0$.

Note: In order to integrate $dz = \dfrac{\partial z}{\partial x}dx + \dfrac{\partial z}{\partial y}dy = pdx + qdy$, we must know p and q in terms of x, y, z .

For this purpose, introduce another non-linear partial differential equation of the first order of the form

$$F(x, y, z, p, q, a) = 0 ,$$

involving an arbitrary constant 'a' compatible with $f(x, y, z, p, q) = 0..$

Solving $f(x, y, z, p, q) = 0.$ and $F(x, y, z, p, q, a) = 0,$, we get

$$p = p(x, y, z, a), q = q(x, y, z, b)$$

On substitution of $p = p(x, y, z, a), q = q(x, y, z, b)$ in $dz = \dfrac{\partial z}{\partial x}dx + \dfrac{\partial z}{\partial y}dy = pdx + qdy$, equation $dz = \dfrac{\partial z}{\partial x}dx + \dfrac{\partial z}{\partial y}dy = pdx + qdy$ becomes integrable, resulting in the complete solution of $f(x, y, z, p, q) = 0.$ in the form

$$\phi(x, y, z, a, b) = 0,$$

containing two arbitrary constants a and b.

To determine F: We differentiate $f(x, y, z, p, q) = 0$ and $F(x, y, z, p, q, a) = 0$, partially w. r. t. x and y. Thus

$$\frac{\partial f}{\partial x} + \frac{\partial f}{\partial z}\cdot p + \frac{\partial f}{\partial p}\cdot\frac{\partial p}{\partial x} + \frac{\partial f}{\partial q}\cdot\frac{\partial q}{\partial x} = 0,$$

$$\frac{\partial F}{\partial x} + \frac{\partial F}{\partial z}\cdot p + \frac{\partial F}{\partial p}\cdot\frac{\partial p}{\partial x} + \frac{\partial F}{\partial q}\cdot\frac{\partial q}{\partial x} = 0,$$

$$\frac{\partial f}{\partial y} + \frac{\partial f}{\partial z} \cdot q + \frac{\partial f}{\partial p} \cdot \frac{\partial p}{\partial y} + \frac{\partial f}{\partial q} \cdot \frac{\partial q}{\partial y} = 0,$$

$$\frac{\partial F}{\partial y} + \frac{\partial F}{\partial z} \cdot q + \frac{\partial F}{\partial p} \cdot \frac{\partial p}{\partial y} + \frac{\partial F}{\partial q} \cdot \frac{\partial q}{\partial y} = 0.$$

Eliminating $\dfrac{\partial p}{\partial x}$ between $\dfrac{\partial f}{\partial x} + \dfrac{\partial f}{\partial z} \cdot p + \dfrac{\partial f}{\partial p} \cdot \dfrac{\partial p}{\partial x} + \dfrac{\partial f}{\partial q} \cdot \dfrac{\partial q}{\partial x} = 0,$ and

$\dfrac{\partial F}{\partial x} + \dfrac{\partial F}{\partial z} \cdot p + \dfrac{\partial F}{\partial p} \cdot \dfrac{\partial p}{\partial x} + \dfrac{\partial F}{\partial q} \cdot \dfrac{\partial q}{\partial x} = 0,$ we get

$$\left(\frac{\partial f}{\partial x} \cdot \frac{\partial F}{\partial p} - \frac{\partial F}{\partial x} \cdot \frac{\partial f}{\partial p} \right) + \left(\frac{\partial f}{\partial z} \cdot \frac{\partial F}{\partial p} - \frac{\partial F}{\partial z} \cdot \frac{\partial f}{\partial p} \right) \cdot p + \left(\frac{\partial f}{\partial q} \cdot \frac{\partial F}{\partial p} - \frac{\partial F}{\partial q} \cdot \frac{\partial f}{\partial p} \right) \frac{\partial q}{\partial x} = 0.$$

Eliminating $\dfrac{\partial q}{\partial y}$ between $\dfrac{\partial f}{\partial y} + \dfrac{\partial f}{\partial z} \cdot q + \dfrac{\partial f}{\partial p} \cdot \dfrac{\partial p}{\partial y} + \dfrac{\partial f}{\partial q} \cdot \dfrac{\partial q}{\partial y} = 0,$ and

$\dfrac{\partial F}{\partial y} + \dfrac{\partial F}{\partial z} \cdot q + \dfrac{\partial F}{\partial p} \cdot \dfrac{\partial p}{\partial y} + \dfrac{\partial F}{\partial q} \cdot \dfrac{\partial q}{\partial y} = 0,$ we get

$$\left(\frac{\partial f}{\partial y} \cdot \frac{\partial F}{\partial q} - \frac{\partial F}{\partial y} \cdot \frac{\partial f}{\partial q} \right) + \left(\frac{\partial f}{\partial z} \cdot \frac{\partial F}{\partial q} - \frac{\partial F}{\partial z} \cdot \frac{\partial f}{\partial q} \right) \cdot q + \left(\frac{\partial f}{\partial q} \cdot \frac{\partial F}{\partial q} - \frac{\partial F}{\partial p} \cdot \frac{\partial f}{\partial q} \right) \frac{\partial p}{\partial y} = 0$$

Since $\dfrac{\partial q}{\partial x} = \dfrac{\partial^2 z}{\partial x \partial y} = \dfrac{\partial^2 z}{\partial y \partial x} = \dfrac{\partial p}{\partial y}$ and the last term in above 2 equation differ in sign only, then adding above 2 equations, we get

$$\left(\frac{\partial f}{\partial x} + p \frac{\partial f}{\partial z} \right) \frac{\partial F}{\partial p} + \left(\frac{\partial f}{\partial y} + q \frac{\partial f}{\partial z} \right) \frac{\partial F}{\partial q} + \left(-p \frac{\partial f}{\partial p} - q \frac{\partial f}{\partial q} \right) \frac{\partial F}{\partial z} + \left(-\frac{\partial f}{\partial p} \right) \frac{\partial F}{\partial x} + \left(-\frac{\partial f}{\partial q} \right) \frac{\partial F}{\partial y} = 0,$$

which is the linear partial differential equation (Lagrange's linear equation) of the first order with x, y, z, p, q as independent variables and F as the dependent variable.

$\therefore$ The auxiliary above equations of

$$\left(\frac{\partial f}{\partial x} + p \frac{\partial f}{\partial z} \right) \frac{\partial F}{\partial p} + \left(\frac{\partial f}{\partial y} + q \frac{\partial f}{\partial z} \right) \frac{\partial F}{\partial q} + \left(-p \frac{\partial f}{\partial p} - q \frac{\partial f}{\partial q} \right) \frac{\partial F}{\partial z} + \left(-\frac{\partial f}{\partial p} \right) \frac{\partial F}{\partial x} + \left(-\frac{\partial f}{\partial q} \right) \frac{\partial F}{\partial y} = 0, \text{ are}$$

$$\frac{dp}{\dfrac{\partial f}{\partial x} + p \dfrac{\partial f}{\partial z}} = \frac{dq}{\dfrac{\partial f}{\partial y} + q \dfrac{\partial f}{\partial z}} = \frac{dz}{-p \dfrac{\partial f}{\partial p} - q \dfrac{\partial f}{\partial q}} = \frac{dx}{-\dfrac{\partial f}{\partial p}} = \frac{dy}{-\dfrac{\partial f}{\partial q}}.$$

These above equations are known as Charpit's equations.

We can also write the above Equation as

$$\frac{dx}{f_p} = \frac{dy}{f_q} = \frac{dz}{pf_p + qf_q} = \frac{dp}{-\left(f_x + pf_z \right)} = \frac{dq}{-\left(f_y + qf_z \right)}$$

Example: Find a complete integral of

$$p^2 x + q^2 y = z.$$

Solution: To find a complete integral, we proceed as follows.

Step 1: (Computing f_x, f_y, f_z, f_p, f_q).

Set $f \equiv p^2 x + q^2 y - z = 0$. Then

$$f_x = p^2, \ f_y = q^2, f_z = -1, f_p = 2px, f_q = 2qy$$

$$\Rightarrow pf_p + qf_q = 2p^2 x + 2q^2 y, \ -(f_x + pf_z) = -p^2 + p, -(f_y + qf_z) = -q^2 + q.$$

Step 2: (Writing Charpit's equations and finding a solution $g(x, y, z, p, q, a)$)

The Charpit's equations (or auxiliary) equations are:

$$\frac{dx}{f_p} = \frac{dy}{f_q} = \frac{dz}{pf_p + qf_q} = \frac{dp}{-(f_x + pf_z)} = \frac{dq}{-(f_y + qf_z)}$$

$$\Rightarrow \frac{dx}{2px} = \frac{dy}{2qy} = \frac{dz}{2(p^2 x + q^2 y)} = \frac{dp}{-p^2 + p} = \frac{dq}{-q^2 + q}$$

From which it follows that

$$\frac{p^2 dx + 2pxdp}{2p^3 x + 2p^2 x - 2p^3 x} = \frac{q^2 dy + 2qydq}{2q^3 y + 2q^2 y - 2q^3 y}$$

$$\Rightarrow \frac{p^2 dx + 2pxdp}{p^2 x} = \frac{q^2 dy + 2qydq}{q^2 y}$$

On integrating, we obtain

$$\log(p^2 x) = \log(q^2 y) + \log a$$

$$\Rightarrow p^2 x = aq^2 y,$$

Where a is an arbitrary constant.

Step 3: (Solving for p and q).

Using $p^2 x + q^2 y = z.$ and $p^2 x = aq^2 y,$, we find that

$$p^2 x + q^2 y = z, \ p^2 x = aq^2 y$$

$$\Rightarrow (aq^2 y) + q^2 y = z \Rightarrow q^2 y(1 + a) = z$$

$$\Rightarrow q^2 = \frac{z}{(1+a)y} \Rightarrow q = \left[\frac{z}{(1+a)y}\right]^{1/2}.$$

and

$$p^2 = aq^2 \frac{y}{x} = a\frac{z}{(1+a)y}\frac{y}{x} = \frac{az}{(1+a)x}$$

$$\Rightarrow p = \left[\frac{az}{(1+a)x}\right]^{1/2}$$

Step 4: (Writing $dz = p(x, y, z, a)dx + q(x, y, z, a)dy$ and finding its solution).

Writing

$$dz = \left[\frac{az}{(1+a)x}\right]^{1/2} dx + \left[\frac{z}{(1+a)y}\right]^{1/2} dy$$

$$\Rightarrow \left(\frac{1+a}{z}\right)^{1/2} dz = \left(\frac{a}{x}\right)^{1/2} dx + \left(\frac{1}{y}\right)^{1/2} dy.$$

Integrate to have

$$\left[(1+a)z\right]^{1/2} = (ax)^{1/2} + (y)^{1/2} + b$$

the complete integral of the equation $p^2 x + q^2 y = z..$

Cauchy Problem

The objective to solve PDE

$$F\left(x, y, z, z_x, z_y\right) = 0$$

Subject to an appropriate initial condition (i.e., z assume prescribed values on some curve).

Let $(f(s), g(s))$ traces out a regular curve in the xy -plane as s varies. We regard this curve as being an initial curve. We seek a solution $u(x, y)$ of the following problem (known as Cauchy's problem).

$$F\left(x, y, z, z_x, z_y\right) = 0, u\left(f(s), g(s)\right) = G(s)$$

Where G(s) is a continuously differentiable function. Such a problem may have no solution (e.g., the PDE $z_x^2 + z_y^2 + 1 = 0$). However, if a solution exists in some neighborhood of the initial curve, then such a solution can often be determined using the following steps.

Step 1: Find functions $h(s)$ and $k(s)$ (if possible) such that

$$F\big(f(s), g(s), G(s), h(s), k(s)\big) = 0, \; G'(s) = h(s)f'(s) + k(s)g'(s) \text{ and}$$

$$F_p\big(f(s), g(s), G(s), h(s), k(s)\big)g'(s) - F_q\big(f(s), g(s), G(s), h(s), k(s)\big)f'(s) \neq 0.$$

Note that if $h(s)$ and $k(s)$ do not exist, then $F\big(x, y, z, z_x, z_y\big) = 0,\; u\big(f(s), g(s)\big) = G(s)$ has no solution.

If there are several choices for $\big(h(s),\, k(s)\big)$, then a solution of $F\big(x, y, z, z_x, z_y\big) = 0,\; u\big(f(s), g(s)\big) = G(s)$ exists for each such choice.

Step 2: For each fixed s, solve the following characteristics system for $x(s, t)$, $y(s, t)$, $z(s, t)$, $p(s, t)$, $q(s, t)$ with the given initial conditions $p(s, 0) = h(s)$, $q(s, 0) = k(s)$, Where $h(s)$ and $k(s)$ are the functions found in Step 1.

$$\frac{d}{dt}x(s, t) = F_p\big(x(s, t), y(s, t), z(s, t), p(s, t), q(s, t)\big)$$

$$\frac{d}{dt}y(s, t) = F_q\big(x(s, t), y(s, t), z(s, t), p(s, t), q(s, t)\big)$$

$$\frac{d}{dt}z(s, t) = p(s, t)F_p\big(x(s, t), y(s, t), z(s, t), p(s, t), q(s, t)\big)$$
$$+ q(s, t)F_q\big(x(s, t), y(s, t), z(s, t), p(s, t), q(s, t)\big)$$

$$\frac{d}{dt}p(s, t) = -[F_x\big(x(s, t), y(s, t), z(s, t), p(s, t), q(s, t)\big)$$
$$+ p(s, t)F_z\big(x(s, t), y(s, t), z(s, t), p(s, t), q(s, t)\big)]$$

$$\frac{d}{dt}q(s, t) = -[F_y\big(x(s, t), y(s, t), z(s, t), p(s, t), q(s, t)\big)$$
$$+ q(s, t)F_z\big(x(s, t), y(s, t), z(s, t), p(s, t), q(s, t)\big)]$$

Step 3: As s and t vary, the point (x, y, z), defined by

$$x = x(s, t), \quad y = y(s, t), \quad z = z(s, t)$$

traces out the graph of a solution z of $F\big(x, y, z, z_x, z_y\big) = 0,\; u\big(f(s), g(s)\big) = G(s)$ in the xyz-space, in a neighborhood of the curve traced out by $\big(f(s), g(s), G(s)\big)$.

In some cases, one can use the first two equations in $x = x(s, t)$, $y = y(s, t)$, $z = z(s, t)$ to solve for s and t in terms of x and y $\big($say, $s = s(x, y)$ and $t = t(x, y)\big)$ to obtain a solution $z(x, y) = z\big(s(x, y), t(x, y)\big)$, for (x, y) in a neighborhood of the curve$\big(f(s), g(s)\big)$.

Example: Solve the PDE $z_x z_y - z = 0$ subject to the condition $z(s, -s) = 1$.

Solution: Here, we have

$$F(x, y, z, p, q) = pq - z.$$

The characteristics system given in Step 2 takes the form

$$\frac{dx}{dt} = F_p = q(t), \quad \frac{dy}{dt} = F_q = p(t), \quad \frac{dz}{dt} = pF_p + qF_q = 2p(t)q(t),$$

$$\frac{dp}{dt} = -\left[F_x + p(t)F_z\right] = p(t), \quad \frac{dq}{dt} = -\left[F_y + q(t)F_z\right] = q(t).$$

Note that

$$\frac{dp}{dt} = p(t) \Rightarrow p(t) = ce^t \text{ and } \frac{dq}{dt} = q(t) \Rightarrow q(t) = de^t,$$

Where c and d are arbitrary constants. Since we are looking for a characteristics strip (i.e., $F(x, y, z, p, q) = 0$), we set $z(t) = p(t)q(t) = cde^{2t}$. The equations for the characteristic strip are:

$$x(t) = de^t + d_1, \quad y(t) = ce^t + c_1, \quad z(t) = cde^{2t}, \quad p(t) = ce^t, \quad q(t) = de^t,$$

Where c_1 and d_1 are constants.

The initial condition $z(s, -s) = 1$ is given on the line $y = -x$ traced out by $(s, -s)$, in $F(x, y, z, z_x, z_y) = 0$, $u(f(s), g(s)) = G(s)$, we have $f(s) = s$ and $g(s) = -s$. We must find $h(s)$ and $k(s)$ such that

$$1 = G(s) = h(s)k(s) \quad 0 = G'(s) = h(s) - k(s),$$

$$0 \neq F_p(\ldots)(-1) - F_q(\ldots)(1) = -k(s) - h(s).$$

Thus, we have two choices $h(s) = 1$ and $k(s) = 1$, or $h(s) = -1$ and $k(s) = -1$. For the choice $h(s) = 1$ and $k(s) = 1$, we obtain

$$x(s, t) = e^t - 1 + s, \quad y(s, t) = e^t - 1 - s, \quad z(s, t) = e^{2t}, \quad p(s, t) = e^t, \quad q(s, t) = e^t.$$

From the first two equations, we obtain

$$e^t = (x + y + 2)/2.$$

Then the solution is

$$z(x, y) = e^{2t} = \frac{(x + y + 2)^2}{4}.$$

If we choose $h(s) = -1$ and $k(s) = -1$, the solution is given by

$$z(x, y) = \frac{(x + y - 2)^2}{4}.$$

First-order Partial Differential Equation

A first order PDE in two independent variables x, y and the dependent variable z can be written in the form

$$f\left(x, y, z, \frac{\partial z}{\partial x}, \frac{\partial z}{\partial y}\right) = 0.$$

For convenience, we set

$$p = \frac{\partial z}{\partial x}, \quad q = \frac{\partial z}{\partial y}.$$

Equation $f\left(x, y, z, \dfrac{\partial z}{\partial x}, \dfrac{\partial z}{\partial y}\right) = 0.$ then takes the form

$$f(x, y, z, p, q) \quad 0.$$

The equations of the type $f(x, y, z, p, q) = 0$ arise in many applications in geometry and physics. For instance, consider the following geometrical problem.

Example: Find all functions $z(x, y)$ such that the tangent plane to the graph $z = z(x, y)$ at any arbitrary point $(x_0, y_0, z(x_0, y_0))$ passes through the origin characterized by the PDE $xz_x + yz_y - z = 0$.

The equation of the tangent plane to the graph at $(x_0, y_0, z(x_0, y_0))$ is

$$z_x(x_0, y_0)(x - x_0) + z_y(x_0, y_0)(y - y_0) - (z - z(x_0, y_0)) = 0.$$

This plane passes through the origin (0, 0, 0) and hence, we must have

$$-z_x(x_0, y_0)\, x_0 - z_y(x_0, y_0) y_0 + z(x_0, y_0) = 0.$$

For the equation $-z_x(x_0, y_0)\, x_0 - z_y(x_0, y_0) y_0 + z(x_0, y_0) = 0.$ to hold for all (x_0, y_0) in the domain of z, z must satisfy

$$xz_x + yz_y - z = 0,$$

which is a first-order PDE.

Example: The set of all spheres with centers on the z-axis is characterized by the first-order PDE $yp - xq = 0$.

The equation

$$x^2 + y^2 + (z - c)^2 = r^2,$$

Where r and c are arbitrary constants, represents the set of all spheres whose centers lie on the z-axis. Differentiating $x^2 + y^2 + (z - c)^2 = r^2$ with respect to x, we obtain

$$2\left(x + (z - c)\frac{\partial z}{\partial x}\right) = 2\left(x + (z - c)p\right) = 0.$$

Differentiate $x^2 + y^2 + (z - c)^2 = r^2$ with respect to y to have

$$y + (z - c)q = 0.$$

Eliminating the arbitrary constant c from $2\left(x + (z - c)\frac{\partial z}{\partial x}\right) = 2\left(x + (z - c)p\right) = 0.$ and $y + (z - c)q = 0$, we obtain the first-order PDE

$$yp - xq = 0$$

Equation $x^2 + y^2 + (z - c)^2 = r^2$ in some sense characterized the first-order PDE $yp - xq = 0$.

Example: Consider all surfaces described by an equation of the form

$$z = f(x^2 + y^2),$$

Where f is an arbitrary function, described by the first-order PDE.

Writing $u = x^2 + y^2$ and differentiating $z = f(x^2 + y^2)$, with respect to x and y, it follows that

$$p = 2x f'(u); \quad q = 2yf'(u)$$

Where $f'(u) = \dfrac{df}{du}$. Eliminating $f'(u)$ from the above two equations, we obtain the same first-order PDE as in $yp - xq = 0$.

Formation of First-order PDEs

The applications of conservation principles often yield a first-order PDEs. We have seen in the previous two examples that a first-order PDE can be formed either by eliminating arbitrary constants or an arbitrary function involved. Below, we now generalize the arguments of above 2 examples to show that how a first-order PDE can be formed.

Method I (Eliminating arbitrary constants): Consider two parameters family of surfaces described by the equation

$$F(x, y, z, a, b) = 0$$

Where a and b are arbitrary constants. Equation $F(x, y, z, a, b) = 0$ may be thought of as a generalization of the relation $x^2 + y^2 + (z - c)^2 = r^2$.

Differentiating $F(x, y, z, a, b) = 0$ with respect to x and y, we obtain

$$\frac{\partial F}{\partial x} + p\,\frac{\partial F}{\partial z} = 0$$

$$\frac{\partial F}{\partial y} + q\,\frac{\partial F}{\partial z} = 0.$$

Eliminate the constants a, b from equations $F(x, y, z, a, b) = 0$, $\frac{\partial F}{\partial x} + p\frac{\partial F}{\partial z} = 0$ and $\frac{\partial F}{\partial y} + q\frac{\partial F}{\partial z} = 0.$ to obtain a first-order PDE of the form

$$f(x, y, z, p, q) = 0.$$

This shows that a family of surfaces described by the relation $F(x, y, z, a, b) = 0$ gives rise to a first-order PDE $f(x, y, z, p, q) = 0$.

Method II (Eliminating arbitrary function): Now consider the generalization of last Example given above Let $u(x, y, z) = c_1$ and $v(x, y, z) = c_2$ be two known functions of x, y and z satisfying a relation of the form

$$F(u, v) = 0,$$

Where F is an arbitrary function of u and v. Differentiating $F(u, v) = 0$, with respect to x and y lead to the equations

$$F_u(u_x + u_z p) + F_v(v_x + v_z p) = 0$$

$$F_u(u_y + u_z q) + F_v(v_y + v_z q) = 0.$$

Eliminating F_u and F_v from the above two equations, we obtain

$$p\,\frac{\partial(u, v)}{\partial(y, z)} + q\,\frac{\partial(u, v)}{\partial(z, x)} = \frac{\partial(u, v)}{\partial(x, y)},$$

which is a first-order PDE of the form $f(x, y, z, p, q) = 0$. Here, $\frac{\partial(u,v)}{\partial(x,y)} = u_x v_y - u_y v_x$.

Classification of First-order PDEs

We classify the equation $f\left(x, y, z, \frac{\partial z}{\partial x}, \frac{\partial z}{\partial y}\right) = 0.$ depending on the special forms of the function

f. If $f\left(x, y, z, \frac{\partial z}{\partial x}, \frac{\partial z}{\partial y}\right) = 0.$ is of the form

$$a(x, y)\frac{\partial z}{\partial x} + b(x, y)\frac{\partial z}{\partial y} + c(x, y)z = d(x, y)$$

then it is called linear first-order PDE. Note that the function f is linear in $\frac{\partial z}{\partial x}$, $\frac{\partial z}{\partial y}$ and z with all coefficients depending on the independent variables x and y only.

If $f\left(x, y, z, \frac{\partial z}{\partial x}, \frac{\partial z}{\partial y}\right) = 0.$ has the form

$$a(x, y)\frac{\partial z}{\partial x} + b(x, y)\frac{\partial z}{\partial y} = c(x, y, z)$$

then it is called semi linear because it is linear in the leading (highest-order) terms $\frac{\partial z}{\partial x}$ and $\frac{\partial z}{\partial y}$. However, it need not be linear in z. Note that the coefficients of $\frac{\partial z}{\partial x}$ and $\frac{\partial z}{\partial y}$ are functions of the independent variables only.

If $f\left(x, y, z, \frac{\partial z}{\partial x}, \frac{\partial z}{\partial y}\right) = 0.$ has the form

$$a(x, y, z)\frac{\partial z}{\partial x} + b(x, y, z)\frac{\partial z}{\partial y} = c(x, y, z)$$

then it is called quasi-linear PDE. Here the function f is linear in the derivatives $\frac{\partial z}{\partial x}$ and $\frac{\partial z}{\partial y}$ with the coefficients a, b and c depending on the independent variables x and y as well as on the unknown z. Note that linear and semi linear equations are special cases of quasi-linear equations.

Any equation that does not fit into one of these forms is called nonlinear.

Second-order Partial Differential Equation

An equation is said to be of order two, if it involves at least one of the differential coefficients $r = (\partial^2 z / \partial^2 x)$, $s = (\partial^2 z / \partial x \partial y)$, $t = (\partial^2 z / \partial^2 y)$, but now of higher order; the quantities p and q may also enter into the equation. Thus the general form of a second order Partial differential equation is

$$f(x, y, z, p, q, r, s, t) = 0$$

The most general linear partial differential equation of order two in two independent variables x and y with variable coefficients is of the form

$$Rr + Ss + Tt + Pp + Qq + Zz = F$$

Where R, S, T, P, Q, Z, F are functions of x and y only and not all R, S, T are zero

Example: Solve $r = 6x$.

Solution: The given equation can be written as $\frac{\partial^2 z}{\partial x^2} = 6x$

Integrating $\dfrac{\partial^2 z}{\partial x^2} = 6x$ w. r. t. x we get $x\dfrac{\partial z}{\partial x} = 3x^2 + \varnothing_1(y)$

Where $\varnothing_1(y)$ is an arbitrary function of y.

Integrating $x\dfrac{\partial z}{\partial x} = 3x^2 + \varnothing_1(y)$ w. r. t. x we get

$$x\ z = x^3 + x\varnothing_1(y) + \varnothing_2(y)$$

Where $\varnothing_2(y)$ is an arbitrary function of y.

Example: $ar = xy$

Solution: Given equation can be written as $\dfrac{\partial^2 z}{\partial x^2} = \dfrac{1}{a}xy$

Integrating $\dfrac{\partial^2 z}{\partial x^2} = \dfrac{1}{a}xy$ w. r. t., x, we get

$$\dfrac{\partial z}{\partial x} = \left(\dfrac{y}{a}\right)\dfrac{x^2}{2} + \varnothing_1(y)$$

Where $\varnothing_1(y)$ is an arbitrary function of y

Integrating $\dfrac{\partial z}{\partial x} = \left(\dfrac{y}{a}\right)\dfrac{x^2}{2} + \varnothing_1(y)$ w. r. t., x,

$$z = \left(\dfrac{y}{a}\right)\dfrac{3}{6} + x\varnothing_1(y) + \varnothing_2(y)$$

Or $z = \dfrac{y}{2a} + x\varnothing_1(y) + \varnothing_2(y)$

Where $\varnothing_2(y)$ is an arbitrary function of y.

Classification of Second order PDE

The classification of PDE is motivated by the classification of second order algebraic equations in two-variables

$$ax^2 + bxy + cy^2 + dx + ey + f = 0.$$

We know that the nature of the curves will be decided by the principal part $ax^2 + bxy + cy^2$ i.e., the term containing highest degree. Depending on the sign of the discriminant $b^2 - 4ac$, we classify the curve as follows:

If $b^2 - 4ac > 0$ then the curve traces hyperbola.

If $b^2 - 4ac = 0$ then the curve traces parabola.

If $b^2 - 4ac < 0$ then the curve traces ellipse.

With suitable transformation, we can transform $ax^2 + bxy + cy^2 + dx + ey + f = 0$ into the following normal form,

$$\frac{x^2}{a^2} - \frac{y^2}{b^2} = 1 \ (\text{hyperbola}).$$

$$x^2 = y \ (\text{parabola})$$

$$\frac{x^2}{a^2} + \frac{y^2}{b^2} = 1 \ (\text{ellipse}).$$

Linear PDE with constant coefficients: Let us first consider the following general linear second order PDE in two independent variables x and y with constant coefficients:

$$Au_{xx} + Bu_{xy} + Cu_{yy} + Du_x + Eu_y + Fu + G = 0$$

Where the coefficients A, B, C, D, E, F and G are constants. The nature of the equation $Au_{xx} + Bu_{xy} + Cu_{yy} + Du_x + Eu_y + Fu + G = 0$ is determined by the principal part containing highest partial derivatives i.e.,

$$Lu \equiv Au_{xx} + Bu_{xy} + Cu_{yy}.$$

For classification, we attach a symbol to $Lu \equiv Au_{xx} + Bu_{xy} + Cu_{yy}.$ as $P(x, y) = Ax^2 + Bxy + Cy^2$ (as if we have replaced x by $\dfrac{\partial}{\partial x}$ and y by $\dfrac{\partial}{\partial y}$). Now depending on the sign of the discriminant $\left(B^2 - 4AC\right)$, the classification of $Au_{xx} + Bu_{xy} + Cu_{yy} + Du_x + Eu_y + Fu + G = 0$ is done as follows:

$$B^2 - 4AC > 0 \Rightarrow \left(Au_{xx} + Bu_{xy} + Cu_{yy} + Du_x + Eu_y + Fu + G = 0\right) \text{ is hyperbolic}$$
$$B^2 - 4AC = 0 \Rightarrow \left(Au_{xx} + Bu_{xy} + Cu_{yy} + Du_x + Eu_y + Fu + G = 0\right) \text{ is parabolic}$$
$$B^2 - 4AC < 0 \Rightarrow \left(Au_{xx} + Bu_{xy} + Cu_{yy} + Du_x + Eu_y + Fu + G = 0\right) \text{ is elliptic}$$

Linear PDE with variable coefficients: The above classification of $Au_{xx} + Bu_{xy} + Cu_{yy} + Du_x + Eu_y + Fu + G = 0$ is still valid if the coefficients A, B, C, D, E and F depend on x, y. In this case, the conditions $B^2 - 4AC > 0 \Rightarrow \left(Au_{xx} + Bu_{xy} + Cu_{yy} + Du_x + Eu_y + Fu + G = 0\right)$ or $B^2 - 4AC = 0 \Rightarrow \left(Au_{xx} + Bu_{xy} + Cu_{yy} + Du_x + Eu_y + Fu + G = 0\right)$ and $B^2 - 4AC < 0 \Rightarrow \left(Au_{xx} + Bu_{xy} + Cu_{yy} + Du_x + Eu_y + Fu + G = 0\right)$ should be satisfied at each point (x, y) in the region Where we want to describe its nature e.g., for elliptic we need to verify

$$B^2(x, y) - 4A(x, y)C(x, y) < 0$$

for each (x, y) in the region of interest. Thus, we classify linear PDE with variable coefficients as follows:

$B^2(x, y) - 4A(x, y)C(x, y) > 0$ at $(x, y) \Rightarrow \left(Au_{xx} + Bu_{xy} + Cu_{yy} + Du_x + Eu_y + Fu + G = 0\right)$ is hyperbolic at (x, y)

$B^2(x, y) - 4A(x, y)C(x, y) = 0$ at $(x, y) \Rightarrow \left(Au_{xx} + Bu_{xy} + Cu_{yy} + Du_x + Eu_y + Fu + G = 0\right)$ is parabolic at (x, y)

$B^2(x, y) - 4A(x, y)C(x, y) < 0$ at $(x, y) \Rightarrow \left(Au_{xx} + Bu_{xy} + Cu_{yy} + Du_x + Eu_y + Fu + G = 0\right)$ is elliptic at (x, y)

Note: Eq. $Au_{xx} + Bu_{xy} + Cu_{yy} + Du_x + Eu_y + Fu + G = 0$ is hyperbolic, parabolic, or elliptic depends only on the coefficients of the second derivatives. It has nothing to do with the first-derivative terms, the term in u, or the nonhomogeneous term.

Classification with More than Two Variables

Consider the second-order PDE in general form:

$$\sum_{i=1}^{n}\sum_{j=1}^{n} a_{ij} \frac{\partial^2 u}{\partial x_i \partial x_j} + \sum_{i=1}^{n} b_i \frac{\partial u}{\partial x_i} + cu + d = 0,$$

Where the coefficients a_{ij}, b_i, c and d are functions of $x = (x_1, x_2, \cdots, x_n)$ alone and

$$u = u(x_1, x_2, \cdots, x_n).$$

Its principal part is

$$L \equiv \sum_{i=1}^{n}\sum_{j=1}^{n} a_{ij} \frac{\partial^2}{\partial x_i \partial x_j}.$$

It is enough to assume that $A = [a_{ij}]$ is symmetric if not, let $\bar{a}_{ij} = \frac{1}{2}(a_{ij} + a_{ji})$ and rewrite

$$L \equiv \sum_{i=1}^{n}\sum_{j=1}^{n} \bar{a}_{ij} \frac{\partial^2}{\partial x_i \partial x_j}$$

Note that $\dfrac{\partial^2 u}{\partial x_i \partial x_j} = \dfrac{\partial^2 u}{\partial x_j \partial x_i}$. As in two-space dimension, let us attach a quadratic form P with

$$L \equiv \sum_{i=1}^{n}\sum_{j=1}^{n} \bar{a}_{ij} \frac{\partial^2}{\partial x_i \partial x_j} \quad \text{(i.e., replacing} \frac{\partial u}{\partial x_i} \text{ by } x_i \text{).}$$

$$P(x_1, x_2, \cdots, x_n) = \sum_{i=1}^{n}\sum_{j=1}^{n} a_{ij}x_i x_j .$$

Since A is a real valued symmetric $(a_{ij} = a_{ji})$ matrix, it is diagonalizable with real $\lambda_1, \lambda_2, \ldots, \lambda_n$ (counted with their multiplicities). In other words, there exists a corresponding set of orthonormal set of n eigenvectors, say $\sigma_1, \sigma_2, \cdots, \sigma_n$ with $R = [\sigma_1, \sigma_2, \cdots, \sigma_n]$ as column vectors such that

$$R \ AR = \begin{bmatrix} & & & \\ & \cdot & & \\ & & \cdot & \\ & & & \cdot \\ & & & \end{bmatrix} = D$$

We now classify ($\sum\limits_{i=1}^{n}\sum\limits_{j=1}^{n} a_{ij} \dfrac{\partial^2 u}{\partial x_i \partial x_j} + \sum\limits_{i=1}^{n} b_i \dfrac{\partial u}{\partial x_i} + cu + d = 0,$) depending on sign of eigenvalues of A:

(a) If $\lambda_i > 0 \; \forall i$ or $\lambda_i < 0 \; \forall i$ then $\sum\limits_{i=1}^{n}\sum\limits_{j=1}^{n} a_{ij} \dfrac{\partial^2 u}{\partial x_i \partial x_j} + \sum\limits_{i=1}^{n} b_i \dfrac{\partial u}{\partial x_i} + cu + d = 0$, is elliptic type.

(b) If one or more of the $\lambda_i = 0$ then $\sum\limits_{i=1}^{n}\sum\limits_{j=1}^{n} a_{ij} \dfrac{\partial^2 u}{\partial x_i \partial x_j} + \sum\limits_{i=1}^{n} b_i \dfrac{\partial u}{\partial x_i} + cu + d = 0$, is parabolic type.

(c) If one of the $\lambda_i < 0$ or $\lambda_i > 0$ and all the remaining have opposite sign then $\sum\limits_{i=1}^{n}\sum\limits_{j=1}^{n} a_{ij} \dfrac{\partial^2 u}{\partial x_i \partial x_j} +$

$\sum\limits_{i=1}^{n} b_i \dfrac{\partial u}{\partial x_i} + cu + d = 0$, is said to be of hyperbolic type.

Hyperbolic Partial Differential Equation

A partial differential equation of second-order, i.e., one of the form

$$A u_{xx} + 2B u_{xy} + C u_{yy} + D u_x + E u_y + F = 0,$$

is called hyperbolic if the matrix

$$Z \equiv \begin{bmatrix} A & B \\ B & C \end{bmatrix}$$

satisfies $\det(Z) < 0$.

Wave Equation

A classical example of a hyperbolic PDE is the wave equation:

$$f_{tt} = c^2 \nabla^2 f$$

The wave equation applies to problems in vibrations, electrostatics, gas dynamics, acoustics, etc. The general features of the wave equation are illustrated in this section for the problem of unsteady one-dimensional acoustic wave propagation.

Fluid flow is governed by the law of conservation of mass (the continuity equation), Newton's second law of motion (the momentum equation), and the first law of thermodynamics (the energy equation).

$$\rho_t + \nabla.(\rho \mathbf{V}) = 0$$
$$\rho \mathbf{V}_t + \rho(\mathbf{V}.\nabla)\mathbf{V} + \nabla P = 0$$
$$P_t + \mathbf{V}.\nabla P - a^2(p_t + \mathbf{V}.\nabla p) = 0$$

Where ρ is the fluid density (kg/m³), V is the fluid velocity vector (m/s), P is the static pressure (N/m^2), and a is the speed of propagation of small disturbances (m/s) (i.e., speed of sound). Above Equations are restricted to the flow of a pure substance with no body forces or transport phenomena (i.e., no mass, momentum, or energy diffusion). For unsteady one-dimensional flow, $\rho_t + \nabla.(\rho \mathbf{V}) = 0$ to $\rho \mathbf{V}_t + \rho(\nabla.\nabla)\mathbf{V} + \nabla P = 0$, $P_t + \mathbf{V}.\nabla P - a^2(p_t + \mathbf{V}.\nabla p) = 0$ yield:

$$\rho_t + p u_x + u p_x = 0$$

$$\rho u_t + \rho u u_x + P_x = 0$$

$$P_t + u P_x - a^2(\rho_t + u\rho_x) = 0$$

Above Equations are more general examples of the simple one-dimensional convection equation

$$f_t + u f_x = 0$$

Where the property f is being convected by the velocity u through the solution domain $D(x, t)$. Equation $f_t + u f_x = 0$ in three independent variables is

$$f_t + u f_x + v f_y + w f_z = f_t + \mathbf{V}.\nabla f = \frac{Df}{Dt} = 0$$

Where $u, v,$ and w are the velocity components in the $x, y,$ and z directions, respectively, and the vector operator D/Dt is called the substantial derivative:

$$\frac{D}{Dt} = \frac{\partial}{\partial t} + u\frac{\partial}{\partial x} + v\frac{\partial}{\partial y} + w\frac{\partial}{\partial z} = \frac{\partial}{\partial t} + \mathbf{V}\cdot\nabla$$

Equations $\rho_t + p u_x + u p_x = 0$, $\rho u_t + \rho u u_x + P_x =$ and $P_t + u P_x - a^2(\rho_t + u\rho_x) = 0$ are frequently combined to eliminate the derivatives of density. Thus,

$$P_t + u P_x + \rho a^2 u_x = 0$$

Equations $\rho_t + p u_x + u p_x = 0$ to $\rho u_t + \rho u u_x + P_x = 0$, $P_t + u P_x - a^2(\rho_t + u\rho_x) = 0$ or Eqs. $\rho u_t + \rho u u_x + P_x = 0$ and $P_t + u P_x + \rho a^2 u_x = 0$, are classical examples system of nonlinear first-order PDEs.

Acoustics is the science devoted to the study of the motion of small amplitude disturbances in a fluid medium. Consider the classical case of infinitesimally small perturbations in velocity, pressure, and density in a stagnant fluid. In that case,

$$u = u_0 + u' = u' \qquad P = P_o + P' \qquad \rho = \rho_0 + \rho' \qquad a = a_0 + a'$$

Where u_0, P_o, ρ_0, and a_0 are the undisturbed properties of the fluid, and u', P', ρ', and a' are infinitesimal perturbations.

For a stagnant fluid, $u_o = 0$. Substituting Eq. ($u = u_0 + u' = u'$ $P = P_o + P'$ $\rho = \rho_0 + \rho'$ $a = a_0 + a'$) into Eqs. ($\rho u_t + \rho u u_x + P_x = 0$) and ($P_t + u P_x + \rho a^2 u_x = 0$) and neglecting all products of perturbation quantities yields following system of linear PDEs:

$$P_o u_t' + P_x' = 0$$

$$P_t' + \rho_0 a_0^2 u_x' = 0$$

Above equations can be combined to solve explicitly for either the pressure perturbation P' or the velocity perturbation u'. Differentiating $P_o u_t' + P_x' = 0$ with respect to x and $P_t' + \rho_0 a_0^2 u_x' = 0$ equation with respect to t and combining the results to eliminate u_{xt}' yields the wave equation for the pressure perturbation, P':

$$P_{tt}' = a_0^2 P_{xx}'$$

Differentiating ($P_o u_t' + P_x' = 0$) with respect to t and ($P_t' + \rho_0 a_0^2 u_x' = 0$) with respect to x combining the results to eliminate P_{xt}' yields the wave equation for the velocity perturbation u':

$$u_{tt}' = a_0^2 u_{xx}'$$

Equations ($P_{tt}' = a_0^2 P_{xx}'$) and ($u_{tt}' = a_0^2 u_{xx}'$) show that the properties of a linearized acoustic field are governed by the wave equation. In terms of the general second-order PDE defined by Equation $A f_{xx} + B f_{xy} + C f_{yy} + D f_x + E f_y + F f = G$, $A = 1$, $B = 0$, and $C = $ -a_0^2. The discriminant, $B^2 - 4AC$, is

$$B^2 - 4AC = 0 - 4(1)\left(-a_0^2\right) = 4a_0^2 > 0$$

Consequently, Eqs. ($P_{tt}' = a_0^2 P_{xx}'$) and ($u_{tt}' = a_0^2 u_{xx}'$) are hyperbolic PDEs

Since ($P_{tt}' = a_0^2 P_{xx}'$) and ($u_{tt}' = a_0^2 u_{xx}'$) both involve the same differential operators [i.e., $(\ )_{tt} = a_0^2 (\ )_{xx}$], they have the same characteristics. Consequently, it is necessary to study only one of them, so Eq. ($P_{tt}' = a_0^2 P_{xx}'$) is chosen. The characteristics associated with Eq. ($P_{tt}' = a_0^2 P_{xx}'$) are determined by performing a characteristics analysis.

In this case, Eq. becomes

$$\begin{bmatrix} A & B & C \\ dx & dy & 0 \\ 0 & dx & dy \end{bmatrix} \begin{bmatrix} f_{xx} \\ f_{xy} \\ f_{yy} \end{bmatrix} = \begin{bmatrix} -Df_x - Ef_y - F + G \\ d(f_x) \\ d(f_y) \end{bmatrix}$$

$$\begin{bmatrix} 1 & 0 & -a_0^2 \\ dt & dx & 0 \\ 0 & dt & dx \end{bmatrix} \begin{bmatrix} P_{tt}' \\ P_{xt}' \\ P_{xx}' \end{bmatrix} = \begin{bmatrix} 0 \\ d(P_x') \\ d(P_t') \end{bmatrix}$$

The characteristic equation corresponding to equation $P_{tt}' = a_0^2 \, P_{xx}'$ is determined by setting

the determinant of the coefficient matrix of Equation $\begin{bmatrix} 1 & 0 & -a_0^2 \\ dt & dx & 0 \\ 0 & dt & dx \end{bmatrix} \begin{bmatrix} P_{tt}' \\ P_{xt}' \\ P_{xx}' \end{bmatrix} = \begin{bmatrix} 0 \\ d(P_x') \\ d(P_t') \end{bmatrix}$

to zero and solving for the slopes the characteristic paths. This yields

$$(dx)^2 - a_0^2 (dt)^2 = 0$$

Above equation is a quadratic equation for dx / dt. Solving for dx / dt gives

$$\frac{dx}{dt} = \pm a_0$$

$$x = x_0 \pm a_0 t$$

Equation ($\frac{dx}{dt} = \pm a_0$) shows that there are two real distinct roots associated with the characteristic equation, and Eq. ($x = x_0 \pm a_0 t$) shows that the characteristic paths are straight lines having the slopes $\pm 1 / a_o$ in the xt plane. The speed of propagation of information along these characteristic paths is

$$c \frac{dx}{dt} = \pm a_0$$

Consequently, information propagates at the acoustic speed a_0 along the characteristic paths. This situation is illustrated schematically in Figure below. Information at point P propagates at a finite rate in physical space. Consequently, the perturbation pressure at point P depends only upon the solution within the finite domain of dependence illustrated in Figure below. Likewise, the perturbation pressure at point P influences the solution only within the finite range of influence illustrated in Figure below .The finite speed of propagation of information and the finite domain of dependence and range of influence must be accounted for when solving hyperbolic PDEs.

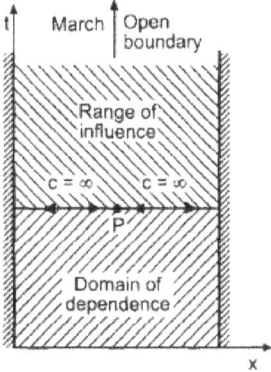

Figure: Solution domain for a parabolic propagation problem

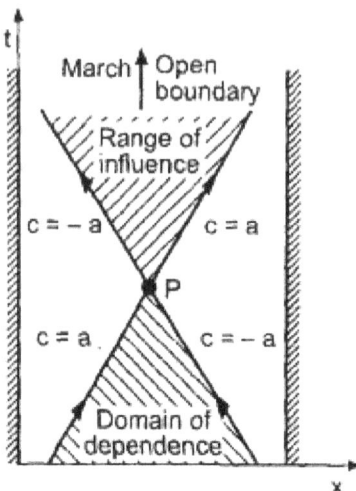

Figure: Solution domain for a hyperbolic propagation problem

Equations ($P_0 u'_t + P'_x = 0$) and ($P'_t + \rho_0 a_0^2 u'_x = 0$) are examples of a system of two coupled first-order convection equations of the general form:

$$f_t + a g_x = 0$$
$$g_t + a f_x = 0$$

Differentiating Eq. ($f_t + a g_x = 0$) with respect to t, differentiating Eq. ($g_t + a f_x = 0$) with respect and multiplying by a, and subtracting yields the wave equation:

$$f_{tt} = a^2 f_{xx}$$

Consequently, the second-order wave equation can be interpreted as a system of two coupled first-order convection equations.

Euler Equations

Euler equations are notably hyperbolic conservation equations in the case without external field (i.e., in the limit of high Froude number). In fact, like any Cauchy equation, the Euler equations originally formulated in convective form (also called usually "Lagrangian form", but this name is not self-explanatory and historically wrong, so it will be avoided) can also be put in the "conservation form" (also called usually "Eulerian form", but also this name is not self-explanatory and is historically wrong, so it will be avoided here). The conservation form emphasizes the mathematical interpretation of the equations as conservation equations through a control volume fixed in space, and is the most important for these equations also from a numerical point of view. The convective form emphasizes changes to the state in a frame of reference moving with the fluid.

Incompressible Euler Equations with Constant and Uniform Density

In convective form (i.e., the form with the convective operator made explicit in the momentum equation), the incompressible Euler equations in case of density constant in time and uniform in space are:

Incompressible Euler equations with constant and uniform density (*convective or Lagrangian form*)

$$\begin{cases} \dfrac{D\mathbf{u}}{Dt} = -\nabla w + \mathbf{g} \\ \quad \nabla \cdot \mathbf{u} = 0 \end{cases}$$

Where:

- $\mathbf{u}$ is the flow velocity vector, with components in an N-dimensional space $u_1, u_2, \ldots, u_N,$,

- $\dfrac{D}{Dt}$ denotes the material derivative in time,

- $\cdot$ denotes the scalar product,

- ∇ is the nabla operator, here used to represent the specific thermodynamic work gradient (first equation), and

- $\nabla \cdot \mathbf{u}$ is the flow velocity divergence (second equation),

- w is the specific (with the sense of *per unit mass*) thermodynamic work, the internal source term.

- $\mathbf{g}$ represents body accelerations (per unit mass) acting on the continuum, for example gravity, inertial accelerations, electric field acceleration, and so on.

The first equation is the Euler momentum equation with uniform density (for this equation it could also not be constant in time). By expanding the material derivative, the equations become:

$$\begin{cases} \dfrac{\partial \mathbf{u}}{\partial t} + \mathbf{u} \cdot \nabla \mathbf{u} = -\nabla w + \mathbf{g} \\ \qquad\quad \nabla \cdot \mathbf{u} = 0 \end{cases}$$

In fact for a flow with uniform density ρ_0 the following identity holds:

$$\nabla w \equiv \nabla \left(\frac{p}{\rho_0} \right) = \frac{1}{\rho_0} \nabla p$$

Where p is the mechanic pressure. The second equation is the incompressible constraint, stating the flow velocity is a solenoidal field (the order of the equations is not casual, but underlines the fact that the incompressible constraint is not a degenerate form of the continuity equation, but rather of the energy equation, as it will become clear in the following). Notably, the continuity equation would be required also in this incompressible case as an additional third equation in case of density varying in time *or* varying in space. For example, with density uniform but varying in time, the continuity equation to be added to the above set would correspond to:

$$\frac{\partial \rho}{\partial t} = 0$$

So the case of constant and uniform density is the only one not requiring the continuity equation as additional equation regardless of the presence or absence of the incompressible constraint. In fact, the case of incompressible Euler equations with constant and uniform density being analyzed is a toy model featuring only two simplified equations, so it is ideal for didactical purposes even if with limited physical relevancy.

The equations above thus represent respectively conservation of mass (1 scalar equation) and momentum (1 vector equation containing N scalar components, Where N is the physical dimension of the space of interest). Flow velocity and pressure are the so-called *physical variables*.

In a coordinate system given by $(x_1,...,x_N)$ the velocity and external force vectors $\mathbf{u}$ and $\mathbf{g}$ have components $(u_1,...,u_N)$ and $(g_1,...,g_N)$,, respectively. Then the equations may be expressed in subscript notation as:

$$\begin{cases} \dfrac{\partial u_i}{\partial t} + \displaystyle\sum_{j=1}^{N} \dfrac{\partial(u_i u_j + w\delta_{ij})}{\partial x_j} = g_i \\ \displaystyle\sum_{i=1}^{N} \dfrac{\partial u_i}{\partial x_i} = 0 \end{cases}$$

Where the i and j subscripts label the N-dimensional space components, and δ_{ij} is the Kroenecker delta. The use of Einstein notation (Where the sum is implied by repeated indices instead of sigma notation) is also frequent.

Properties

Although Euler first presented these equations in 1755, many fundamental questions about them remain unanswered.

In three space dimensions it is not even known whether solutions of the equations are defined for all time or if they form singularities.

Smooth solutions of the free (in the sense of without source term: g=0) equations satisfy the conservation of specific kinetic energy:

$$\frac{\partial}{\partial t}\left(\frac{1}{2}u^2\right) + \nabla \cdot (u^2\mathbf{u} + w\mathbf{I}) = 0$$

In the one dimensional case without the source term (both pressure gradient and external force), the momentum equation becomes the inviscid Burgers equation:

$$\frac{\partial u}{\partial t} + u\frac{\partial u}{\partial x} = 0$$

This is a model equation giving many insights on Euler equations.

Nondimensionalisation

In order to make the equations dimensionless, a characteristic length r_0, and a characteristic velocity u_0, need to be defined. These should be chosen such that the dimensionless variables are all of order one. The following dimensionless variables are thus obtained:

$$u^* \equiv \frac{u}{u_0},$$

$$r^* \equiv \frac{r}{r_0},$$

$$t^* \equiv \frac{u_0}{r_0}t,$$

$$p^* \equiv \frac{w}{u_0^2},$$

$$\nabla^* \equiv r_0\nabla$$

And of the field unit vector:

$$\hat{g} \equiv \frac{g}{g},$$

Substitution of these inversed relations in Euler equations, defining the Froude number, yields (omitting the * at apix):

Incompressible Euler equations with constant and uniform density (*nondimensional form*)

$$\begin{cases} \dfrac{D\mathbf{u}}{Dt} = -\nabla w + \dfrac{1}{Fr}\hat{g} \\ \nabla \cdot \mathbf{u} = 0 \end{cases}$$

Euler equations in the Froude limit (no external field) are named free equations and are conservative. The limit of high Froude numbers (low external field) is thus notable and can be studied with perturbation theory.

Conservation Form

The conservation form emphasizes the mathematical properties of Euler equations, and especially the contracted form is often the most convenient one for computational fluid dynamics simulations. Computationally, there are some advantages in using the conserved variables. This gives rise to a large class of numerical methods called conservative methods.

The free Euler equations are conservative, in the sense they are equivalent to a conservation equation:

$$\frac{\partial \mathbf{y}}{\partial t} + \nabla \cdot \mathbf{F} = 0,$$

or simply in Einstein notation:

$$\frac{\partial y_j}{\partial t} + \frac{\partial f_{ij}}{\partial r_i} = 0_i,$$

Where the conservation quantity $\mathbf{y}$ in this case is a vector, and $\mathbf{F}$ is a flux matrix. This can be simply proved.

Demonstration of the Conservation Form

First, the following identities hold:

$$\nabla \cdot (w\mathbf{I}) = \mathbf{I} \cdot \nabla w + w\nabla \cdot \mathbf{I} = \nabla w$$

$$\mathbf{u} \cdot \nabla \cdot \mathbf{u} = \nabla \cdot (\mathbf{u} \otimes \mathbf{u})$$

Where $\otimes$ denotes the outer product. The same identities expressed in Einstein notation are:

$$\partial_i (w\delta_{ij}) = \delta_{ij}\partial_i w + w\partial_i\delta_{ij} = \delta_{ij}\partial_i w = \partial_j w$$

$$u_j\partial_j u_i = \partial_i (u_i u_j)$$

Where I is the identity matrix with dimension N and δ_{ij} its general element, the Kroenecker delta.

Thanks to these vector identities, the incompressible Euler equations with constant and uniform density and without external field can be put in the so-called *conservation*(or Eulerian) differential form, with vector notation:

$$\begin{cases} \dfrac{\partial \mathbf{u}}{\partial t} + \nabla \cdot (\mathbf{u} \otimes \mathbf{u} + w\mathbf{I}) = \mathbf{0} \\ \dfrac{\partial 0}{\partial t} + \nabla \cdot \mathbf{u} = 0, \end{cases}$$

or with Einstein notation:

$$\begin{cases} \partial_t u_j + \partial_i (u_i u_j + w\delta_{ij}) = 0 \\ \partial_t 0 + \partial_j u_j = 0, \end{cases}$$

Then incompressible Euler equations with uniform density have conservation variables:

$$\mathbf{y} = \begin{pmatrix} \mathbf{u} \\ 0 \end{pmatrix}; \qquad \mathbf{F} = \begin{pmatrix} \mathbf{u} \otimes \mathbf{u} + w\mathbf{I} \\ \mathbf{u} \end{pmatrix}.$$

Note that in the second component u is by itself a vector, with length N, so y has length N+1 and F has size N(N+1). In 3D for example y has length 4, I has size 3x3 and F has size 4x3, so the explicit forms are:

$$\mathbf{y} = \begin{pmatrix} u_1 \\ u_2 \\ u_3 \\ 0 \end{pmatrix}; \quad \mathbf{F} = \begin{pmatrix} u_1^2 + w & u_1 u_2 & u_1 u_3 \\ u_2 u_1 & u_2^2 + w & u_2 u_3 \\ u_3 u_1 & u_3 u_2 & u_3^2 + w \\ u_1 & u_2 & u_3 \end{pmatrix}.$$

At last Euler equations can be recast into the particular equation:

Incompressible Euler equation(s) with constant and uniform density (*conservation or Eulerian form*)

$$
\frac{\partial}{\partial t}\begin{pmatrix} \mathbf{u} \\ 0 \end{pmatrix} + \nabla \cdot \begin{pmatrix} \mathbf{u} \otimes \mathbf{u} + w\mathbf{I} \\ \mathbf{u} \end{pmatrix} = \begin{pmatrix} \mathbf{g} \\ 0 \end{pmatrix}
$$

Spatial Dimensions

For certain problems, especially when used to analyze compressible flow in a duct or in case the flow is cylindrically or spherically symmetric, the one-dimensional Euler equations are a useful first approximation. Generally, the Euler equations are solved by Riemann's method of characteristics. This involves finding curves in plane of independent variables (i.e., x and t) along which partial differential equations (PDE's) degenerate into ordinary differential equations (ODE's). Numerical solutions of the Euler equations rely heavily on the method of characteristics.

Incompressible Euler equations

In convective form the incompressible Euler equations in case of density variable in space are:

Incompressible Euler equations (*convective or Lagrangian form*)

$$
\begin{cases}
\dfrac{D\rho}{Dt} = 0 \\[2mm]
\dfrac{D\mathbf{u}}{Dt} = -\dfrac{\nabla p}{\rho} + \mathbf{g} \\[2mm]
\nabla \cdot \mathbf{u} = 0
\end{cases}
$$

Where the additional variables are:

* ρ is the fluid mass density,

* p is the pressure, $p = \rho w$.

The first equation, which is the new one, is the incompressible continuity equation. In fact the general continuity equation would be:

$$
\frac{\partial \rho}{\partial t} + \mathbf{u} \cdot \nabla \rho + \rho \nabla \cdot \mathbf{u} = 0
$$

but here the last term is identically zero for the incompressibility constraint.

Conservation form

The incompressible Euler equations in the Froude limit are equivalent to a single conservation equation with conserved quantity and associated flux respectively:

$$\mathbf{y} = \begin{pmatrix} \rho \\ \rho\mathbf{u} \\ 0 \end{pmatrix}; \qquad \mathbf{F} = \begin{pmatrix} \rho\mathbf{u} \\ \rho\mathbf{u}\otimes\mathbf{u} + p\mathbf{I} \\ \mathbf{u} \end{pmatrix}.$$

Here $\mathbf{y}$ has length $N+2$ and $\mathbf{F}$ has size $(N+2)\times N$. In general (not only in the Froude limit) Euler equations are expressible as:

$$\frac{\partial}{\partial t}\begin{pmatrix} \rho \\ \rho\mathbf{u} \\ 0 \end{pmatrix} + \nabla\cdot\begin{pmatrix} \rho\mathbf{u} \\ \rho\mathbf{u}\otimes\mathbf{u} + p\mathbf{I} \\ \mathbf{u} \end{pmatrix} = \begin{pmatrix} 0 \\ \rho\mathbf{g} \\ 0 \end{pmatrix}$$

Conservation Variables

The variables for the equations in conservation form are not yet optimised. In fact we could define:

$$\mathbf{y} = \begin{pmatrix} \rho \\ \mathbf{j} \\ 0 \end{pmatrix}; \qquad \mathbf{F} = \begin{pmatrix} \mathbf{j} \\ \dfrac{1}{\rho}\mathbf{j}\otimes\mathbf{j} + p\mathbf{I} \\ \dfrac{\mathbf{j}}{\rho} \end{pmatrix}.$$

Where:

$\mathbf{j} = \rho\mathbf{u}$ is the momentum density, a conservation variable.

Incompressible Euler equation(s) (*conservation or Eulerian form*)

$$\frac{\partial}{\partial t}\begin{pmatrix} \rho \\ \mathbf{j} \\ 0 \end{pmatrix} + \nabla\cdot\begin{pmatrix} \mathbf{j} \\ \dfrac{1}{\rho}\mathbf{j}\otimes\mathbf{j} + p\mathbf{I} \\ \dfrac{\mathbf{j}}{\rho} \end{pmatrix} = \begin{pmatrix} 0 \\ \mathbf{f} \\ 0 \end{pmatrix}$$

Where:

$\mathbf{f} = \rho\mathbf{g}$ is the force density, a conservation variable.

Euler Equations

In differential convective form, the compressible (and most general) Euler equations can be written shortly with the material derivative notation:

Euler equations (*convective form*)

$$\begin{cases} \dfrac{D\rho}{Dt} = -\rho \nabla \cdot \mathbf{u} \\[2mm] \dfrac{D\mathbf{u}}{Dt} = -\dfrac{\nabla p}{\rho} + \mathbf{g} \\[2mm] \dfrac{De}{Dt} = -\dfrac{p}{\rho} \nabla \cdot \mathbf{u} \end{cases}$$

Where the additional variables here is:

- e is the specific internal energy (internal energy per unit mass).

The equations above thus represent conservation of mass, momentum, and energy: the energy equation expressed in the variable internal energy allows to understand the link with the incompressible case, but it is not in the simplest form. Mass density, flow velocity and pressure are the so-called *convective variables* (or physical variables, or lagrangian variables), while mass density, momentum density and total energy density are the so-called *conserved variables* (also called eulerian, or mathematical variables). If one explicitates the material derivative the equations above are:

$$\begin{cases} \dfrac{\partial \rho}{\partial t} + \mathbf{u} \cdot \nabla \rho + \rho \nabla \cdot \mathbf{u} = 0 \\[2mm] \dfrac{\partial \mathbf{u}}{\partial t} + \mathbf{u} \cdot \nabla \mathbf{u} + \dfrac{\nabla p}{\rho} = \mathbf{g} \\[2mm] \dfrac{\partial e}{\partial t} + \mathbf{u} \cdot \nabla e + \dfrac{p}{\rho} \nabla \cdot \mathbf{u} = 0 \end{cases}$$

Incompressible Constraint

Coming back to the incompressible case, it now becomes apparent that the *incompressible constraint* typical of the former cases actually is a particular form valid for incompressible flows of the *energy equation*, and not of the mass equation. In particular, the incompressible constraint corresponds to the following very simple energy equation:

$$\frac{De}{Dt} = 0$$

Thus for an incompressible inviscid fluid the specific internal energy is constant along the flow lines, also in a time-dependent flow. The pressure in an incompressible flow acts like a Lagrange multiplier, being the multiplier of the incompressible constraint in the energy equation, and consequently in incompressible flows it has no thermodynamic meaning. In fact, thermodynamics is typical of compressible flows and degenerates in incompressible flows. Basing on the mass conservation equation, one can put this equation in the conservation form:

$$\frac{\partial \rho e}{\partial t} + \nabla \cdot (\rho e \mathbf{u}) = 0$$

meaning that for an incompressible inviscid nonconductive flow a continuity equation holds for the internal energy.

Enthalpy Conservation

Since by definition the specific enthalpy is:

$$h = e + \frac{p}{\rho}$$

The material derivative of the specific internal energy can be expressed as:

$$\frac{De}{Dt} = \frac{Dh}{Dt} - \frac{1}{\rho}\left(\frac{Dp}{Dt} - \frac{p}{\rho}\frac{D\rho}{Dt}\right)$$

Then by substituting the momentum equation in this expression, one obtains:

$$\frac{De}{Dt} = \frac{Dh}{Dt} - \frac{1}{\rho}\left(p\nabla \cdot \mathbf{u} + \frac{Dp}{Dt}\right)$$

And by substituting the latter in the energy equation, one obtains that the enthalpy expression for the Euler energy equation:

$$\frac{Dh}{Dt} = \frac{1}{\rho}\frac{Dp}{Dt}$$

In a reference frame moving with an inviscid and nonconductive flow, the variation of enthalpy directly corresponds to a variation of pressure.

Thermodynamic Systems

In thermodynamics the independent variables are the specific volume, and the specific entropy, while the specific energy is a function of state of these two variables.

Deduction of the form Valid for Thermodynamic Systems

Considering the first equation, variable must be changed from density to specific volume. By definition:

$$v \equiv \frac{1}{\rho}$$

Thus the following identities hold:

$$\nabla \rho = \nabla\left(\frac{1}{v}\right) = -\frac{1}{v^2}\nabla v$$

$$\frac{\partial \rho}{\partial t} = \frac{\partial}{\partial t}\left(\frac{1}{v}\right) = -\frac{1}{v^2}\frac{\partial v}{\partial t}$$

Then by substituting these expressions in the mass conservation equation:

$$-\frac{\mathbf{u}}{v^2}\cdot\nabla v - \frac{1}{v^2}\frac{\partial v}{\partial t} = -\frac{1}{v}\nabla\cdot\mathbf{u}$$

And by multiplication:

$$\frac{\partial v}{\partial t} + \mathbf{u}\cdot\nabla v = v\nabla\cdot\mathbf{u}$$

Note that this equation is the only belonging to general continuum equations, so only this equation have the same form for example also in Navier-Stokes equations.

On the other hand, the pressure in thermodynamics is the opposite of the partial derivative of the specific internal energy with respect to the specific volume:

$$p(v,s) = -\frac{\partial e(v,s)}{\partial v}$$

since the internal energy in thermodynamics is a function of the two variables aforementioned, the pressure gradient contained into the momentum equation should be explicited as:

$$-\nabla p(v,s) = -\frac{\partial p}{\partial v}\nabla v - \frac{\partial p}{\partial s}\nabla s = \frac{\partial^2 e}{\partial v^2}\nabla v + \frac{\partial^2 e}{\partial v\partial s}\nabla s$$

It is convenient for brevity to switch the notation for the second order derivatives:

$$-\nabla p(v,s) = e_{vv}\nabla v + e_{vs}\nabla s$$

Finally, the energy equation:

$$\frac{De}{Dt} = -pv\nabla\cdot\mathbf{u}$$

can be furtherly simplified in convective form by changing variable from specific energy to the specific entropy: in fact the first law of thermodynamics in local form can be written:

$$\frac{De}{Dt} = T\frac{Ds}{Dt} - p\frac{Dv}{Dt}$$

by substituting the material derivative of the internal energy, the energy equation becomes:

$$T\frac{Ds}{Dt} + \frac{p}{\rho^2}\left(\frac{D\rho}{Dt} + \rho\nabla\cdot\mathbf{u}\right) = 0$$

now the term between parenthesis is identically zero according to the conservation of mass, then the Euler energy equation becomes simply:

$$\frac{Ds}{Dt} = 0$$

For a thermodynamic fluid, the compressible Euler equations are consequently best written as:

Euler equations (*convective form, for a thermodynamic system*)

$$\begin{cases} \dfrac{Dv}{Dt} = v\nabla \cdot \mathbf{u} \\[2mm] \dfrac{D\mathbf{u}}{Dt} = ve_{vv}\nabla v + ve_{vs}\nabla s + \mathbf{g} \\[2mm] \dfrac{Ds}{Dt} = 0 \end{cases}$$

Where:

- v is the specific volume

- $\mathbf{u}$ is the flow velocity vector

- s is the specific entropy

Note that, in the general case and not only in the incompressible case, the energy equation means that for an inviscid thermodynamic fluid the specific entropy is constant along the flow lines, also in a time-dependent flow. Basing on the mass conservation equation, one can put this equation in the conservation form:

$$\frac{\partial \rho s}{\partial t} + \nabla \cdot (\rho s \mathbf{u}) = 0$$

meaning that for an inviscid nonconductive flow a continuity equation holds for the entropy.

On the other hand, the two second-order partial derivatives of the specific internal energy in the momentum equation require the specification of the fundamental equation of state of the material considered, i.e. of the specific internal energy as function of the two variables specific volume and specific entropy:

$$e = e(v,s)$$

Note that the *fundamental* equation of state contains all the thermodynamic information about the system, exactly like the couple of a *thermal* equation of state together with a *caloric* equation of state.

Conservation form

The Euler equations in the Froude limit are equivalent to a single conservation equation with conserved quantity and associated flux respectively:

$$y = \begin{pmatrix} \rho \\ \mathbf{j} \\ E^t \end{pmatrix}; \qquad F = \begin{pmatrix} \mathbf{j} \\ \dfrac{1}{\rho}\mathbf{j}\otimes\mathbf{j}+p\mathbf{I} \\ (E^t+p)\dfrac{\mathbf{j}}{\rho} \end{pmatrix}.$$

Where:

- $\mathbf{j}=\rho\mathbf{u}$ is the momentum density, a conservation variable.

- $E^t = \rho e + 1/2\rho u^2$ is the total energy density (total energy per unit volume).

Here y has length N+2 and F has size N(N+2). In general (not only in the Froude limit) Euler equations are expressible as:

Euler equation(s) (*original conservation or Eulerian form*)

$$\frac{\partial}{\partial t}\begin{pmatrix} \rho \\ \mathbf{j} \\ E^t \end{pmatrix} + \nabla\cdot\begin{pmatrix} \mathbf{j} \\ \dfrac{1}{\rho}\mathbf{j}\otimes\mathbf{j}+p\mathbf{I} \\ (E^t+p)\dfrac{\mathbf{j}}{\rho} \end{pmatrix} = \begin{pmatrix} 0 \\ \mathbf{f} \\ \dfrac{\mathbf{j}}{\rho}\cdot\mathbf{f} \end{pmatrix}$$

Where:

- $\mathbf{f}=\rho\mathbf{g}$ is the force density, a conservation variable.

We remark that also the Euler equation even when conservative (no external field, Froude limit) have no Riemann invariants in general. Some further assumptions are required.

However, we already mentioned that for a thermodynamic fluid the equation for the total energy density is equivalent to the conservation equation:

$$\frac{\partial}{\partial t}(\rho s) + \nabla\cdot(\rho s\mathbf{u}) = 0$$

Then the conservation equations in the case of a thermodynamic fluid are more simply expressed as:

Euler equation(s) (conservation form, for thermodynamic fluids)

$$\frac{\partial}{\partial t}\begin{pmatrix} \rho \\ \mathbf{j} \\ S \end{pmatrix} + \nabla\cdot\begin{pmatrix} \mathbf{j} \\ \dfrac{1}{\rho}\mathbf{j}\otimes\mathbf{j}+p\mathbf{I} \\ S\dfrac{\mathbf{j}}{\rho} \end{pmatrix} = \begin{pmatrix} 0 \\ \mathbf{f} \\ 0 \end{pmatrix}$$

Where:

- $S = \rho s$ is the entropy density, a thermodynamic conservation variable.

Another possible form for the energy equation, being particularly useful for isobarics, is:

$$\frac{\partial H'}{\partial t} + \nabla \cdot (H' \mathbf{u}) = \mathbf{u} \cdot \mathbf{f} - \frac{\partial p}{\partial t}$$

Where:

- $H' = E' + p = \rho e + p + 1/2\rho u^2$ is the total enthalpy density.

Quasilinear form and Characteristic Equations

Expanding the fluxes can be an important part of constructing numerical solvers, for example by exploiting (approximate) solutions to the Riemann problem. In regions Where the state vector y varies smoothly, the equations in conservative form can be put in quasilinear form:

$$\frac{\partial \mathbf{y}}{\partial t} + \mathbf{A}_i \frac{\partial \mathbf{y}}{\partial r_i} = \mathbf{0}.$$

Where $\mathbf{A}_i$ are called the flux Jacobians defined as the matrices:

$$\mathbf{A}_i(\mathbf{y}) = \frac{\partial \mathbf{f}_i(\mathbf{y})}{\partial \mathbf{y}}.$$

Obviously this Jacobian does not exist in discontinuity regions (e.g. contact discontinuities, shock waves in inviscid nonconductive flows). Note that if the flux Jacobians $\mathbf{A}_i$ are not functions of the state vector $\mathbf{y}$, the equations reveals *linear*.

Characteristic Equations

The compressible Euler equations can be decoupled into a set of N+2 wave equations that describes sound in Eulerian continuum if they are expressed in characteristic variables instead of conserved variables.

In fact the tensor A is always diagonalizable. If the eigenvalues (the case of Euler equations) are all real the system is defined *hyperbolic*, and physically eigenvalues represent the speeds of propagation of information. If they are all distinguished, the system is defined *strictly hyperbolic* (it will be proved to be the case of one-dimensional Euler equations). Furthermore, note that diagonalisation of compressible Euler equation is easier when the energy equation is expressed in the variable entropy (i.e. with equations for thermodynamic fluids) than in other energy variables. This will become clear by considering the 1D case.

If $\mathbf{p}_i$ is the right eigenvector of the matrix $\mathbf{A}$ corresponding to the eigenvalue λ_i, by building the projection matrix:

$$\mathbf{P} = [\mathbf{p}_1, \mathbf{p}_2, ..., \mathbf{p}_n]$$

One can finally find the *characteristic variables* as:

$$\mathbf{w} = \mathbf{P}^{-1}\mathbf{y},$$

Since A is constant, multiplying the original 1-D equation in flux-Jacobian form with $\mathbf{P}^{-1}$ yields the characteristic equations:

$$\frac{\partial w_i}{\partial t} + \lambda_j \frac{\partial w_i}{\partial r_j} = 0_i$$

The original equations have been decoupled into N+2 characteristic equations each describing a simple wave, with the eigenvalues being the wave speeds. The variables w_i are called the *characteristic variables* and are a subset of the conservative variables. The solution of the initial value problem in terms of characteristic variables is finally very simple. In one spatial dimension it is:

$$w_i(x,t) = w_i(x - \lambda_i t, 0)$$

Then the solution in terms of the original conservative variables is obtained by transforming back:

$$\mathbf{y} = \mathbf{P}\mathbf{w}$$

This computation can be explicated as the linear combination of the eigenvectors:

$$\mathbf{y}(x,t) = \sum_{i=1}^{m} w_i(x - \lambda_i t, 0)\mathbf{p}_i,$$

Now it becomes apparent that the characteristic variables act as weights in the linear combination of the jacobian eigenvectors. The solution can be seen as superposition of waves, each of which is advected independently without change in shape. Each $i - th$ wave has shape $w_i p_i$ and speed of propagation λ_i.

Parabolic Partial Differential Equation

A partial differential equation of second-order, i.e., one of the form

$$A u_{xx} + 2B u_{xy} + C u_{yy} + D u_x + E u_y + F = 0,$$

is called parabolic if the matrix

$$Z \equiv \begin{bmatrix} A & B \\ B & C \end{bmatrix}$$

Satisfies $det(Z) = 0$.

Heat Equation

The heat equation is a parabolic partial differential equation that describes the distribution of heat (or variation in temperature) in a given region over time.

Statement of the Equation

For a function $u(x,y,z,t)$ of three spatial variables (x,y,z) and the time variable t, the heat equation is

$$\frac{\partial u}{\partial t} - \alpha\left(\frac{\partial^2 u}{\partial x^2} + \frac{\partial^2 u}{\partial y^2} + \frac{\partial^2 u}{\partial z^2}\right) = 0$$

More generally in any coordinate system:

$$\frac{\partial u}{\partial t} - \alpha\nabla^2 u = 0$$

Where α is a positive constant, and Δ or ∇^2 denotes the Laplace operator. In the physical problem of temperature variation, $u(x,y,z,t)$ is the temperature and a is the thermal diffusivity. For the mathematical treatment it is sufficient to consider the case $\alpha = 1$.

Note that the state equation, given by the first law of thermodynamics (i.e. conservation of energy), is written in the following form (assuming no mass transfer or radiation). This form is more general and particularly useful to recognize which property (e.g. c_p or ρ) influences which term.

$$\rho c_p \frac{\partial T}{\partial t} - \nabla \cdot (k\nabla T) = \dot{q}_V$$

Where $\dot{q}_V$ is the volumetric heat source.

The heat equation is of fundamental importance in diverse scientific fields. In mathematics, it is the prototypical parabolic partial differential equation. In probability theory, the heat equation is connected with the study of Brownian motion via the Fokker–Planck equation. In financial mathematics it is used to solve the Black–Scholes partial differential equation. The diffusion equation, a more general version of the heat equation, arises in connection with the study of chemical diffusion and other related processes.

General Description

Suppose one has a function u that describes the temperature at a given location (x, y, z). This function will change over time as heat spreads throughout space. The heat equation is used to determine the change in the function u over time. The rate of change of u is proportional to the "curvature" of u. Thus, the sharper the corner, the faster it is rounded off. Over time, the tendency is for peaks to be eroded, and valleys filled in. If u is linear in space (or has a constant gradient) at a given point, then u has reached steady-state and is unchanging at this point (assuming a constant thermal conductivity).

The image to the right is animated and describes the way heat changes in time along a metal bar. One of the interesting properties of the heat equation is the maximum principle that says that the maximum value of u is either earlier in time than the region of concern or on the edge of the region of concern. This is essentially saying that temperature comes either from some source or from earlier in time because heat permeates but is not created from nothingness. This is a property of parabolic partial differential equations and is not difficult to prove mathematically.

Another interesting property is that even if u has a discontinuity at an initial time $t = t_0$, the temperature becomes smooth as soon as $t > t_0$. For example, if a bar of metal has temperature 0 and another has temperature 100 and they are stuck together end to end, then very quickly the temperature at the point of connection will become 50 and the graph of the temperature will run smoothly from 0 to 50.

The heat equation is used in probability and describes random walks. It is also applied in financial mathematics for this reason.

It is also important in Riemannian geometry and thus topology: it was adapted by Richard S. Hamilton when he defined the Ricci flow that was later used by Grigori Perelman to solve the topological Poincaré conjecture.

The Physical Problem and the Equation

Derivation in one Dimension

The heat equation is derived from Fourier's law and conservation of energy By Fourier's law, the rate of flow of heat energy per unit area through a surface is proportional to the negative temperature gradient across the surface,

$$\mathbf{q} = -k\nabla u$$

Where k is the thermal conductivity and u is the temperature. In one dimension, the gradient is an ordinary spatial derivative, and so Fourier's law is

$$q = -k\frac{\partial u}{\partial x}$$

In the absence of work done, a change in internal energy per unit volume in the material, ΔQ, is proportional to the change in temperature, Δu (Δ is the ordinary difference operator with respect to time, not the Laplacian with respect to space). That is,

$$\Delta Q = c_p \rho \Delta u$$

Where c_p is the specific heat capacity and ρ is the mass density of the material. Choosing zero energy at absolute zero temperature, this can be rewritten as,

$$Q = c_p \rho u.$$

The increase in internal energy in a small spatial region of the material.

$$x - \Delta x \leq \xi \leq x + \Delta x$$

over the time period

$$t - \Delta t \leq \tau \leq t + \Delta t$$

is given by

$$c_p \rho \int_{x-\Delta x}^{x+\Delta x} [u(\xi, t+\Delta t) - u(\xi,\, t+\Delta t)]\, d\xi = c_p \rho \int_{t-\Delta t}^{t+\Delta t} \int_{x-\Delta x}^{x+\Delta x} \frac{\partial u}{\partial \tau}\, d\xi\, d\tau$$

Where the fundamental theorem of calculus was used. If no work is done and there are neither heat sources nor sinks, the change in internal energy in the interval $[x-\Delta x,\, x+\Delta x]$ is accounted for entirely by the flux of heat across the boundaries. By Fourier's law, this is

$$k \int_{t-\Delta t}^{t+\Delta t} \left[\frac{\partial u}{\partial x}(x+\Delta x, \tau) - \frac{\partial u}{\partial x}(x-\Delta x, \tau) \right] d\tau = k \int_{t-\Delta t}^{t+\Delta t} \int_{x-\Delta x}^{x+\Delta x} \frac{\partial^2 u}{\partial \xi^2}\, d\xi\, d\tau$$

again by the fundamental theorem of calculus. By conservation of energy,

$$\int_{t-\Delta t}^{t+\Delta t} \int_{x-\Delta x}^{x+\Delta x} [c_p \rho u_\tau - k u_{\xi\xi}]\, d\xi\, d\tau = 0.$$

This is true for any rectangle $[t-\Delta t, t+\Delta t] \times [x-\Delta x, x+\Delta x]$. By the fundamental lemma of the calculus of variations, the integrand must vanish identically:

$$c_p \rho u_t - k u_{xx} = 0.$$

Which can be rewritten as:

$$u_t = \frac{k}{c_p \rho} u_{xx},$$

or:

$$\frac{\partial u}{\partial t} = \frac{k}{c_p \rho} \left(\frac{\partial^2 u}{\partial x^2} \right)$$

which is the heat equation, Where the coefficient (often denoted α)

$$\alpha = \frac{k}{c_p \rho}$$

is called the thermal diffusivity.

An additional term may be introduced into the equation to account for radiative loss of heat, which depends upon the excess temperature $u = T - T_s$ at a given point compared with the surroundings. At low excess temperatures, the radiative loss is approximately μu, giving a one-dimensional heat-transfer equation of the form

$$\frac{\partial u}{\partial t} = \frac{k}{c_p \rho} \left(\frac{\partial^2 u}{\partial x^2} \right) - \mu u.$$

At high excess temperatures, however, the Stefan–Boltzmann law gives a net radiative heat-loss proportional to $T^4 - T_s^4$, and the above equation is inaccurate. For large excess temperatures,

$T^4 - T_s^4 \approx u^4$, giving a high-temperature heat-transfer equation of the form

$$\frac{\partial u}{\partial t} = \alpha \left(\frac{\partial^2 u}{\partial x^2} \right) - mu^4$$

Where $m = \epsilon \sigma p / \rho A c_p$. Here, σ is Stefan's constant, ε is a characteristic constant of the material, p is the sectional perimeter of the bar and A is its cross-sectional area. However, using T instead of u gives a better approximation in this case.

Three-dimensional Problem

In the special cases of wave propagation of heat in an isotropic and homogeneous medium in a 3-dimensional space, this equation is

$$\frac{\partial u}{\partial t} = \alpha \nabla^2 u = \alpha \left(\frac{\partial^2 u}{\partial x^2} + \frac{\partial^2 u}{\partial y^2} + \frac{\partial^2 u}{\partial z^2} \right)$$
$$= \alpha (u_{xx} + u_{yy} + u_{zz})$$

Where:

- $u = u(x, y, z, t)$ is temperature as a function of space and time;
- $\frac{\partial u}{\partial t}$ is the rate of change of temperature at a point over time;
- u_{xx}, u_{yy}, and u_{zz} are the second spatial derivatives (*thermal conductions*) of temperature in the x, y, and z directions, respectively;
- $\alpha = \dfrac{k}{c_p \rho}$ is the thermal diffusivity, a material-specific quantity depending on the *thermal conductivity* k, the *mass density* ρ, and the *specific heat capacity* c_p.

The heat equation is a consequence of Fourier's law of conduction.

If the medium is not the whole space, in order to solve the heat equation uniquely we also need to specify boundary conditions for u. To determine uniqueness of solutions in the whole space it is necessary to assume an exponential bound on the growth of solutions.

Solutions of the heat equation are characterized by a gradual smoothing of the initial temperature distribution by the flow of heat from warmer to colder areas of an object. Generally, many different states and starting conditions will tend toward the same stable equilibrium. As a consequence, to reverse the solution and conclude something about earlier times or initial conditions from the present heat distribution is very inaccurate except over the shortest of time periods.

The heat equation is the prototypical example of a parabolic partial differential equation.

Using the Laplace operator, the heat equation can be simplified, and generalized to similar equations over spaces of arbitrary number of dimensions, as

$$u_t = \alpha \nabla^2 u = \alpha \Delta u,$$

Where the Laplace operator, Δ or ∇^2, the divergence of the gradient, is taken in the spatial variables.

The heat equation governs heat diffusion, as well as other diffusive processes, such as particle diffusion or the propagation of action potential in nerve cells. Although they are not diffusive in nature, some quantum mechanics problems are also governed by a mathematical analog of the heat equation. It also can be used to model some phenomena arising in finance, like the Black–Scholes or the Ornstein-Uhlenbeck processes. The equation, and various non-linear analogues, has also been used in image analysis.

The heat equation is, technically, in violation of special relativity, because its solutions involve instantaneous propagation of a disturbance. The part of the disturbance outside the forward light cone can usually be safely neglected, but if it is necessary to develop a reasonable speed for the transmission of heat, a hyperbolic problem should be considered instead – like a partial differential equation involving a second-order time derivative. Some models of nonlinear heat conduction (which are also parabolic equations) have solutions with finite heat transmission speed.

Internal Heat Generation

The function u above represents temperature of a body. Alternatively, it is sometimes convenient to change units and represent u as the heat density of a medium. Since heat density is proportional to temperature in a homogeneous medium, the heat equation is still obeyed in the new units.

Suppose that a body obeys the heat equation and, in addition, generates its own heat per unit volume (e.g., in watts/litre - W/L) at a rate given by a known function q varying in space and time. Then the heat per unit volume u satisfies an equation

$$\frac{\partial u}{\partial t} = \alpha \left(\frac{\partial^2 u}{\partial x^2} + \frac{\partial^2 u}{\partial y^2} + \frac{\partial^2 u}{\partial z^2} \right) + \frac{1}{c_p \rho} q.$$

For example, a tungsten light bulb filament generates heat, so it would have a positive nonzero value for q when turned on. While the light is turned off, the value of q for the tungsten filament would be zero.

Solving the Heat Equation using Fourier Series

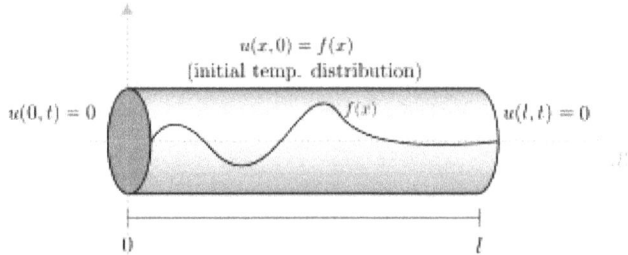

$u(x,0) = f(x)$
(initial temp. distribution)

$u(0,t) = 0$ $f(x)$ $u(l,t) = 0$

0 l

Idealized physical setting for heat conduction in a rod with homogeneous boundary conditions.

The following solution technique for the heat equation was proposed by Joseph Fourier. Let us consider the heat equation for one space variable. This could be used to model heat conduction in a rod. The equation is

$$u_t = \alpha u_{xx}$$

Where $u = u(x, t)$ is a function of two variables x and t. Here

- x is the space variable, so $x \in [0, L]$, Where L is the length of the rod.

- t is the time variable, so $t \geq 0$.

We assume the initial condition

$$u(x,0) = f(x) \quad \forall x \in [0,L]$$

Where the function f is given, and the boundary conditions

$$u(0,t) = 0 = u(L,t) \quad \forall t > 0.$$

Let us attempt to find a solution of $u_t = \alpha u_{xx}$ that is not identically zero satisfying the boundary conditions $u(0,t) = 0 = u(L,t) \quad \forall t > 0$ but with the following property: u is a product in which the dependence of u on x, t is separated, that is:

$$u(x,t) = X(x)T(t).$$

This solution technique is called separation of variables. Substituting u back into equation $u_t = \alpha u_{xx}$,

$$\frac{T'(t)}{\alpha T(t)} = \frac{X''(x)}{X(x)}.$$

Since the right hand side depends only on x and the left hand side only on t, both sides are equal to some constant value $-\lambda$. Thus:

$$T'(t) = -\lambda \alpha T(t)$$

and

$$X''(x) = -\lambda X(x).$$

We will now show that nontrivial solutions for $X''(x) = -\lambda X(x)$ values of $\lambda \leq 0$ cannot occur:

- Suppose that $\lambda < 0$. Then there exist real numbers B, C such that

  $$X(x) = Be^{\sqrt{-\lambda}x} + Ce^{-\sqrt{-\lambda}x}.$$

  From $u(0,t) = 0 = u(L,t) \quad \forall t > 0$. we get $X(0) = 0 = X(L)$ and therefore $B = 0 = C$ which implies u is identically 0.

- Suppose that $\lambda = 0$. Then there exist real numbers B, C such that $X(x) = Bx + C$. From equation $u(0,t) = 0 = u(L,t) \quad \forall t > 0$. we conclude in the same manner as in 1 that u is identically 0.

- Therefore, it must be the case that $\lambda > 0$. Then there exist real numbers A, B, C such that

$$T(t) = Ae^{-\lambda \alpha t}$$

and

$$X(x) = B\sin(\sqrt{\lambda}\,x) + C\cos(\sqrt{\lambda}\,x)$$

From $u(0,t) = 0 = u(L,t)$ $\forall t > 0$. we get $C = 0$ and that for some positive integer n,

$$\sqrt{\lambda} = n\frac{\pi}{L}.$$

This solves the heat equation in the special case that the dependence of u has the special form $u(x,t) = X(x)T(t)$. In general, the sum of solutions to $u_t = \alpha u_{xx}$ that satisfy the boundary conditions $u(0,t) = 0 = u(L,t)$ $\forall t > 0$. also satisfies $u_t = \alpha u_{xx}$ and $u(0,t) = 0 = u(L,t)$ $\forall t > 0$. We can show that the solution to $u_t = \alpha u_{xx}$, $u(x,0) = f(x)$ $\forall x \in [0,L]$ and $u(0,t) = 0 = u(L,t)$ $\forall t > 0$. is given by

$$u(x,t) = \sum_{n=1}^{\infty} D_n \sin\left(\frac{n\pi x}{L}\right) e^{-\frac{n^2\pi^2\alpha t}{L^2}}$$

Where

$$D_n = \frac{2}{L}\int_0^L f(x)\sin\left(\frac{n\pi x}{L}\right)dx.$$

Generalizing the Solution Technique

The solution technique used above can be greatly extended to many other types of equations. The idea is that the operator u_{xx} with the zero boundary conditions can be represented in terms of its eigenvectors. This leads naturally to one of the basic ideas of the spectral theory of linear self-adjoint operators.

Consider the linear operator $\Delta u = u_{xx}$. The infinite sequence of functions

$$e_n(x) = \sqrt{\frac{2}{L}}\sin\left(\frac{n\pi x}{L}\right)$$

for $n \geq 1$ are eigenvectors of Δ. Indeed,

$$\Delta e_n = -\frac{n^2\pi^2}{L^2}e_n.$$

Moreover, any eigenvector f of Δ with the boundary conditions $f(0) = f(L) = 0$ is of the form e_n for some $n \geq 1$. The functions e_n for $n \geq 1$ form an orthonormal sequence with respect to a certain inner product on the space of real-valued functions on $[0, L]$. This means

$$\langle e_n, e_m \rangle = \int_0^L e_n(x) e_m^*(x) dx = \delta_{mn}$$

Finally, the sequence $\{e_n\}_{n \in \mathbf{N}}$ spans a dense linear subspace of $L^2((0, L))$. This shows that in effect we have diagonalized the operator Δ.

Heat Conduction in Non-homogeneous Anisotropic Media

In general, the study of heat conduction is based on several principles. Heat flow is a form of energy flow, and as such it is meaningful to speak of the time rate of flow of heat into a region of space.

- The time rate of heat flow into a region V is given by a time-dependent quantity $q_t(V)$. We assume q has a density Q, so that

$$q_t(V) = \int_V Q(x,t) dx$$

- Heat flow is a time-dependent vector function H(x) characterized as follows: the time rate of heat flowing through an infinitesimal surface element with area dS and with unit normal vector n is

$$\mathbf{H}(x) \cdot \mathbf{n}(x) dS$$

Thus the rate of heat flow into V is also given by the surface integral

$$q_t(V) = -\int_{\partial V} \mathbf{H}(x) \cdot \mathbf{n}(x) dS$$

Where n(x) is the outward pointing normal vector at x.

- The Fourier law states that heat energy flow has the following linear dependence on the temperature gradient

$$\mathbf{H}(x) = -\mathbf{A}(x) \cdot \nabla u(x)$$

Where A(x) is a 3×3 real matrix that is symmetric and positive definite.

- By the divergence theorem, the previous surface integral for heat flow into V can be transformed into the volume integral

$$q_t(V) = -\int_{\partial V} \mathbf{H}(x) \cdot \mathbf{n}(x) dS$$
$$= \int_{\partial V} \mathbf{A}(x) \cdot \nabla u(x) \cdot \mathbf{n}(x) dS$$
$$= \int_V \sum_{i,j} \partial_{x_i} \left(a_{ij}(x) \partial_{x_j} u(x,t) \right) dx$$

- The time rate of temperature change at x is proportional to the heat flowing into an infinitesimal volume element, Where the constant of proportionality is dependent on a constant κ

$$\partial_t u(x,t) = \kappa(x) Q(x,t)$$

Putting these equations together gives the general equation of heat flow:

$$\partial_t u(x,t) = \kappa(x) \sum_{i,j} \partial_{x_i} \left(a_{ij}(x) \partial_{x_j} u(x,t) \right)$$

- The coefficient $\kappa(x)$ is the inverse of specific heat of the substance at $x \times$ density of the substance at x: $\kappa = 1/(\rho c_p)$.

- In the case of an isotropic medium, the matrix A is a scalar matrix equal to thermal conductivity k.

- In the anisotropic case Where the coefficient matrix A is not scalar and/or if it depends on x, then an explicit formula for the solution of the heat equation can seldom be written down. Though, it is usually possible to consider the associated abstract Cauchy problem and show that it is a well-posed problem and/or to show some qualitative properties (like preservation of positive initial data, infinite speed of propagation, convergence toward an equilibrium, smoothing properties). This is usually done by one-parameter semigroups theory: for instance, if A is a symmetric matrix, then the elliptic operator defined by is self-adjoint and dissipative, thus by the spectral theorem it generates a one-parameter semigroup.

$$Au(x) := \sum_{i,j} \partial_{x_i} a_{ij}(x) \partial_{x_j} u(x)$$

Fundamental Solutions

A fundamental solution, also called a *heat kernel*, is a solution of the heat equation corresponding to the initial condition of an initial point source of heat at a known position. These can be used to find a general solution of the heat equation over certain domains; see, for instance, for an introductory treatment.

In one variable, the Green's function is a solution of the initial value problem

$$\begin{cases} u_t(x,t) - k u_{xx}(x,t) = 0 & (x,t) \in \mathbf{R} \times (0,\infty) \\ u(x,0) = \delta(x) \end{cases}$$

Where δ is the Dirac delta function. The solution to this problem is the fundamental solution

$$\Phi(x,t) = \frac{1}{\sqrt{4\pi k t}} \exp\left(-\frac{x^2}{4kt}\right).$$

One can obtain the general solution of the one variable heat equation with initial condition $u(x,0) = g(x)$ for $-\infty < x < \infty$ and $0 < t < \infty$ by applying a convolution:

$$u(x,t) = \int \Phi(x-y,t) g(y) dy.$$

In several spatial variables, the fundamental solution solves the analogous problem

$$\begin{cases} u_t(\mathbf{x},t) - k\sum_{i=1}^{n} u_{x_i x_i}(\mathbf{x},t) = 0 & (\mathbf{x},t) \in \mathbf{R}^n \times (0,\infty) \\ u(\mathbf{x},0) = \delta(\mathbf{x}) \end{cases}$$

The n-variable fundamental solution is the product of the fundamental solutions in each variable; i.e.,

$$\Phi(\mathbf{x},t) = \Phi(x_1,t)\Phi(x_2,t)...\Phi(x_n,t) = \frac{1}{\sqrt{(4\pi kt)^n}} \exp\left(-\frac{\mathbf{x}\cdot\mathbf{x}}{4kt}\right).$$

The general solution of the heat equation on $\mathbf{R}^n$ is then obtained by a convolution, so that to solve the initial value problem with $u(x, 0) = g(x)$, one has

$$u(\mathbf{x},t) = \int_{\mathbf{R}^n} \Phi(\mathbf{x}-\mathbf{y},t)g(\mathbf{y})d\mathbf{y}.$$

The general problem on a domain Ω in $\mathbf{R}^n$ is

$$\begin{cases} u_t(\mathbf{x},t) - k\sum_{i=1}^{n} u_{x_i x_i}(\mathbf{x},t) = 0 & (\mathbf{x},t) \in \Omega \times (0,\infty) \\ u(\mathbf{x},0) = g(\mathbf{x}) & \mathbf{x} \in \Omega \end{cases}$$

with either Dirichlet or Neumann boundary data. A Green's function always exists, but unless the domain Ω can be readily decomposed into one-variable, it may not be possible to write it down explicitly. Other methods for obtaining Green's functions include the method of images, separation of variables, and Laplace transforms.

Some Green's Function Solutions in 1D

A variety of elementary Green's function solutions in one-dimension are recorded here; many others are available elsewhere. In some of these, the spatial domain is $(-\infty,\infty)$. In others, it is the semi-infinite interval $(0,\infty)$ with either Neumann or Dirichlet boundary conditions. One further variation is that some of these solve the inhomogeneous equation

$$u_t = ku_{xx} + f.$$

Where f is some given function of x and t.

Homogeneous Heat Equation

Initial value problem on $(-\infty,\infty)$

$$\begin{cases} u_t = ku_{xx} & (x,t) \in \mathbf{R} \times (0,\infty) \\ u(x,0) = g(x) & IC \end{cases}$$

$$u(x,t) = \frac{1}{\sqrt{4\pi kt}} \int_{-\infty}^{\infty} \exp\left(-\frac{(x-y)^2}{4kt}\right) g(y)dy$$

This solution is the convolution with respect to the variable x of the fundamental solution

$$\Phi(x,t) := \frac{1}{\sqrt{4\pi kt}} \exp\left(-\frac{x^2}{4kt}\right),$$

and the function $g(x)$. (The Green's function number of the fundamental solution is X00.)

Therefore, according to the general properties of the convolution with respect to differentiation, $u = g * \Phi$ is a solution of the same heat equation, for

$$\left(\partial_t - k\partial_x^2\right)(\Phi * g) = \left[\left(\partial_t - k\partial_x^2\right)\Phi\right] * g = 0.$$

Moreover,

$$\Phi(x,t) = \frac{1}{\sqrt{t}} \Phi\left(\frac{x}{\sqrt{t}}\right)$$

$$\int_{-\infty}^{\infty} \Phi(x,t)dx = 1,$$

so that, by general facts about approximation to the identity $\Phi(\cdot,t) * g \rightarrow g$ as $t \rightarrow 0$ in various senses, according to the specific g. For instance, if g is assumed bounded and continuous on R then $(\cdot,t) * g$ converges uniformly to g as $t \rightarrow 0$, meaning that $u(x, t)$ is continuous on $\mathbf{R} \times [0, \infty)$ with $u(x, 0) = g(x)$.

Initial value problem on (0,∞) with homogeneous Dirichlet boundary conditions

$$\begin{cases} u_t = ku_{xx} & (x,t) \in [0,\infty) \times (0,\infty) \\ u(x,0) = g(x) & IC \\ u(0,t) = 0 & BC \end{cases}$$

$$u(x,t) = \frac{1}{\sqrt{4\pi kt}} \int_0^{\infty} \left[\exp\left(-\frac{(x-y)^2}{4kt}\right) - \exp\left(-\frac{(x+y)^2}{4kt}\right)\right] g(y)dy$$

This solution is obtained from the preceding formula as applied to the data $g(x)$ suitably extended to R, so as to be an odd function, that is, letting $g(-x) := -g(x)$ for all x. Correspondingly, the solution of the initial value problem on $(-\infty,\infty)$ is an odd function with respect to the variable x for all values of t, and in particular it satisfies the homogeneous Dirichlet boundary conditions $u(0, t) = 0$. The Green's function number of this solution is X10.

Initial value problem on (0,∞) with homogeneous Neumann boundary conditions

$$\begin{cases} u_t = ku_{xx} & (x,t) \in [0,\infty) \times (0,\infty) \\ u(x,0) = g(x) & IC \\ u_x(0,t) = 0 & BC \end{cases}$$

$$u(x,t) = \frac{1}{\sqrt{4\pi kt}} \int_0^\infty \left[\exp\left(-\frac{(x-y)^2}{4kt}\right) + \exp\left(-\frac{(x+y)^2}{4kt}\right) \right] g(y)\,dy$$

Comment. This solution is obtained from the first solution formula as applied to the data $g(x)$ suitably extended to R so as to be an even function, that is, letting $g(-x) := g(x)$ for all x. Correspondingly, the solution of the initial value problem on R is an even function with respect to the variable x for all values of $t > 0$, and in particular, being smooth, it satisfies the homogeneous Neumann boundary conditions $u_x(0, t) = 0$. The Green's function number of this solution is X20.

Problem on (0,∞) with homogeneous initial conditions and non-homogeneous Dirichlet boundary conditions

$$\begin{cases} u_t = ku_{xx} & (x,t) \in [0,\infty) \times (0,\infty) \\ u(x,0) = 0 & IC \\ u(0,t) = h(t) & BC \end{cases}$$

$$u(x,t) = \int_0^t \frac{x}{\sqrt{4\pi k(t-s)^3}} \exp\left(-\frac{x^2}{4k(t-s)}\right) h(s)\,ds, \qquad \forall x > 0$$

Comment. This solution is the convolution with respect to the variable t of

$$\psi(x,t) := -2k\partial_x \Phi(x,t) = \frac{x}{\sqrt{4\pi kt^3}} \exp\left(-\frac{x^2}{4kt}\right)$$

and the function $h(t)$. Since $\Phi(x,t)$ is the fundamental solution of

$$\partial_t - k\partial_x^2,$$

the function $\psi(x, t)$ is also a solution of the same heat equation, and so is $u := \psi * h$, thanks to general properties of the convolution with respect to differentiation. Moreover,

$$\psi(x,t) = \frac{1}{x^2}\psi\left(1,\frac{t}{x^2}\right)$$

$$\int_0^\infty \psi(x,t)\,dt = 1,$$

so that, by general facts about approximation to the identity, $\psi(x,\cdot) * h \to h$ as $x \to 0$ in various senses, according to the specific h. For instance, if h is assumed continuous on R with support in $[0, \infty)$ then $\psi(x,\cdot) * h$ converges uniformly on compacta to h as $x \to 0$, meaning that $u(x, t)$ is continuous on $[0, \infty) \times [0, \infty)$ with $u(0,t) = h(t)$.

Inhomogeneous Heat Equation

Problem on (-∞,∞) homogeneous initial conditions

$$\begin{cases} u_t = ku_{xx} + f(x,t) & (x,t) \in \mathbf{R} \times (0,\infty) \\ u(x,0) = 0 & IC \end{cases}$$

$$u(x,t) = \int_0^t \int_{-\infty}^\infty \frac{1}{\sqrt{4\pi k(t-s)}} \exp\left(-\frac{(x-y)^2}{4k(t-s)}\right) f(y,s)\,dy\,ds$$

This solution is the convolution in R², that is with respect to both the variables x and t, of the fundamental solution

$$\Phi(x,t) := \frac{1}{\sqrt{4\pi kt}} \exp\left(-\frac{x^2}{4kt}\right)$$

and the function $f(x, t)$, both meant as defined on the whole R² and identically 0 for all $t \to 0$. One verifies that

$$\left(\partial_t - k\partial_x^2\right)(\Phi * f) = f,$$

which expressed in the language of distributions becomes

$$\left(\partial_t - k\partial_x^2\right)\Phi = \delta,$$

Where the distribution δ is the Dirac's delta function, that is the evaluation at 0.

Problem on (0,∞) with Homogeneous Dirichlet Boundary Conditions and Initial conditions

$$\begin{cases} u_t = ku_{xx} + f(x,t) & (x,t) \in [0,\infty) \times (0,\infty) \\ u(x,0) = 0 & IC \\ u(0,t) = 0 & BC \end{cases}$$

$$u(x,t) = \int_0^t \int_0^\infty \frac{1}{\sqrt{4\pi k(t-s)}} \left(\exp\left(-\frac{(x-y)^2}{4k(t-s)}\right) - \exp\left(-\frac{(x+y)^2}{4k(t-s)}\right)\right) f(y,s)\,dy\,ds$$

This solution is obtained from the preceding formula as applied to the data $f(x, t)$ suitably extended to R × [0,∞), so as to be an odd function of the variable x, that is, letting $f(-x, t) := -f(x, t)$ for all x and t. Correspondingly, the solution of the inhomogeneous problem on (−∞,∞) is an odd function with respect to the variable x for all values of t, and in particular it satisfies the homogeneous Dirichlet boundary conditions $u(0, t) = 0$.

Problem on (0,∞) with Homogeneous Neumann Boundary Conditions and Initial Conditions

$$\begin{cases} u_t = ku_{xx} + f(x,t) & (x,t) \in [0,\infty) \times (0,\infty) \\ u(x,0) = 0 & IC \\ u_x(0,t) = 0 & BC \end{cases}$$

$$u(x,t) = \int_0^t \int_0^\infty \frac{1}{\sqrt{4\pi k(t-s)}} \left[\exp\left(-\frac{(x-y)^2}{4k(t-s)}\right) + \exp\left(-\frac{(x+y)^2}{4k(t-s)}\right) \right] f(y,s) \, dy \, ds$$

This solution is obtained from the first formula as applied to the data $f(x, t)$ suitably extended to R × [0,∞), so as to be an even function of the variable x, that is, letting $f(-x, t) := f(x, t)$ for all x and t. Correspondingly, the solution of the inhomogeneous problem on (−∞,∞) is an even function with respect to the variable x for all values of t, and in particular, being a smooth function, it satisfies the homogeneous Neumann boundary conditions $u_x(0, t) = 0$.

Examples

Since the heat equation is linear, solutions of other combinations of boundary conditions, inhomogeneous term, and initial conditions can be found by taking an appropriate linear combination of the above Green's function solutions.

For example, to solve

$$\begin{cases} u_t = ku_{xx} + f & (x,t) \in \mathbf{R} \times (0,\infty) \\ u(x,0) = g(x) & IC \end{cases}$$

let $u = w + v$ Where w and v solve the problems

$$\begin{cases} v_t = kv_{xx} + f, w_t = kw_{xx} & (x,t) \in \mathbf{R} \times (0,\infty) \\ v(x,0) = 0, w(x,0) = g(x) & IC \end{cases}$$

Similarly, to solve

$$\begin{cases} u_t = ku_{xx} + f & (x,t) \in [0,\infty) \times (0,\infty) \\ u(x,0) = g(x) & IC \\ u(0,t) = h(t) & BC \end{cases}$$

let $u = w + v + r$ Where w, v, and r solve the problems

$$\begin{cases} v_t = kv_{xx} + f, w_t = kw_{xx}, r_t = kr_{xx} & (x,t) \in [0,\infty) \times (0,\infty) \\ v(x,0) = 0, w(x,0) = g(x), r(x,0) = 0 & IC \\ v(0,t) = 0, w(0,t) = 0, r(0,t) = h(t) & BC \end{cases}$$

Mean-value Property for the Heat Equation

Solutions of the heat equations

$$(\partial_t - \Delta)u = 0$$

satisfy a mean-value property analogous to the mean-value properties of harmonic functions, solutions of

$$\Delta u = 0,$$

though a bit more complicated. Precisely, if u solves

$$(\partial_t - \Delta)u = 0$$

and

$$(x,t) + E_\lambda \subset \mathrm{dom}(u)$$

then

$$u(x,t) = \frac{\lambda}{4} \int_{E_\lambda} u(x-y,t-s)\frac{|y|^2}{s^2}\, ds\, dy,$$

Where E_λ is a "heat-ball", that is a super-level set of the fundamental solution of the heat equation:

$$E_\lambda := \{(y,s): \Phi(y,s) > \lambda\},$$

$$\Phi(x,t) := (4t\pi)^{-\frac{n}{2}} \exp\left(-\frac{|x|^2}{4t}\right).$$

Notice that

$$\mathrm{diam}(E_\lambda) = o(1)$$

as $\lambda \to \infty$ so the above formula holds for any (x, t) in the (open) set $\mathrm{dom}(u)$ for λ large enough. This can be shown by an argument similar to the analogous one for harmonic functions.

Steady-state Heat Equation

The steady-state heat equation is by definition not dependent on time. In other words, it is assumed conditions exist such that:

$$\frac{\partial u}{\partial t} = 0$$

This condition depends on the time constant and the amount of time passed since boundary conditions have been imposed. Thus, the condition is fulfilled in situations in which the *time equilibrium constant is fast enough* that the more complex time-dependent heat equation can be approximated by the steady-state case. Equivalently, the steady-state condition exists for all cases in which *enough time has passed* that the thermal field u no longer evolves in time.

In the steady-state case, a spatial thermal gradient may (or may not) exist, but if it does, it does not change in time. This equation therefore describes the end result in all thermal problems in which a source is switched on (for example, an engine started in an automobile), and enough time has passed for all permanent temperature gradients to establish themselves in space, after which these spatial gradients no longer change in time (as again, with an automobile in which the engine has been running for long enough). The other (trivial) solution is for all spatial temperature gradients to disappear as well, in which case the temperature become uniform in space, as well.

The equation is much simpler and can help to understand better the physics of the materials without focusing on the dynamic of the heat transport process. It is widely used for simple engineering problems assuming there is equilibrium of the temperature fields and heat transport, with time.

Steady-state condition:

$$\frac{\partial u}{\partial t} = 0$$

The steady-state heat equation for a volume that contains a heat source (the inhomogeneous case), is the Poisson's equation:

$$-k\nabla^2 u = q$$

Where u is the temperature, k is the thermal conductivity and q the heat-flux density of the source.

In electrostatics, this is equivalent to the case Where the space under consideration contains an electrical charge.

The steady-state heat equation without a heat source within the volume (the homogeneous case) is the equation in electrostatics for a volume of free space that does not contain a charge. It is described by Laplace's equation:

$$\nabla^2 u = 0$$

Elliptic Partial Differential Equation

Elliptic equation is any of a class of partial differential equations describing phenomena that do not change from moment to moment, as when a flow of heat or fluid takes place within a medium with no accumulations. The Laplace equation, $u_{xx} + u_{yy} = 0$, is the simplest such equation describing this condition in two dimensions. In addition to satisfying a differential equation within the region, the elliptic equation is also determined by its values (boundary values) along the boundary of the region, which represent the effect from outside the region. These conditions can be either those of a fixed temperature distribution at points of the boundary (Dirichlet problem) or those in which heat is being supplied or removed across the boundary in such a way as to maintain a constant temperature distribution throughout (Neumann problem).

If the highest-order terms of a second-order partial differential equation with constant coefficients are linear and if the coefficients a, b, c of the u_{xx}, u_{xy}, u_{yy} terms satisfy the inequality $b^2 - 4ac < 0$, then, by a change of coordinates, the principal part (highest-order terms) can be written as the Laplacian $u_{xx} + u_{yy}$. Because the properties of a physical system are independent of the coordinate system used to formulate the problem, it is expected that the properties of the solutions of these elliptic equations should be similar to the properties of the solutions of Laplace's equation. If the coefficients a, b, and c are not constant but depend on x and y, then the equation is called elliptic in a given region if $b^2 - 4ac < 0$ at all points in the region. The functions $x^2 - y^2$ and $e^x \cos y$ satisfy the Laplace equation, but the solutions to this equation are usually more complicated because of the boundary conditions that must be satisfied as well.

Laplace's Equation

A classical example of an elliptic PDE is the Laplace equation:

$$\nabla^2 f = 0$$

The Laplace equation applies to problems in ideal fluid flow, mass diffusion, heat diffusion, electrostatics, etc. In the following discussion, the general features of the Laplace equation are illustrated for the problem of steady two-dimensional heat diffusion in a solid.

Consider the differential cube of solid material illustrated in below Figure. Heat flow in a solid is governed by Fourier's law of conduction, which states that

$$\dot{q} = -kA\frac{dT}{dn}$$

Where $\dot{q}$ is the energy transfer per unit time (J/s), T is the temperature (K), A is the across which the energy flows (m^2), dT/dn is the temperature gradient normal to the area A (K/m), and k is the thermal conductivity of the solid (J/m-s-K), which is a physical property of the solid material. The net rate of flow of energy into the solid in the x direction is

$$\dot{q}_{Net, x} = \dot{q}(x) - \dot{q}(x + dx) = \dot{q}(x) - \left[\dot{q}(x) + \frac{\partial \dot{q}(x)}{\partial x} dx\right] = -\frac{\partial \dot{q}(x)}{\partial x}$$

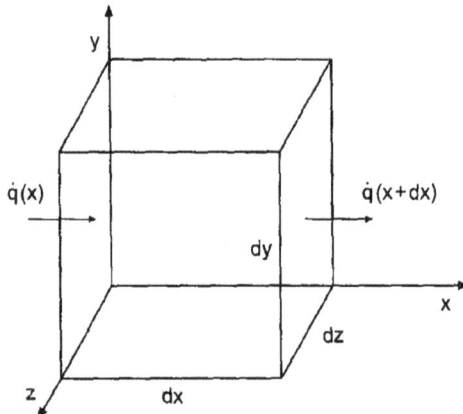

Figure: Physical model of heat diffusion

Introducing Eq. $\dot{q} = -kA\dfrac{dT}{dn}$ into Eq. $\dot{q}_{Net,x} = \dot{q}(x) - \dot{q}(x+dx) = \dot{q}(x) - \left[\dot{q}(x) + \dfrac{\partial \dot{q}(x)}{\partial x}dx\right] = \dfrac{\partial \dot{q}(x)}{\partial x}$
yields

$$\dot{q}_{Net,x} = -\frac{\partial}{\partial x}\left(-kA\frac{\partial T}{\partial x}\right)dx = \frac{\partial}{\partial x}\left(k\frac{\partial T}{\partial x}\right)dV$$

Where $dV = A\,dx$ is the volume of differential cube of solid material. Similarly,

$$\dot{q}_{Net,y} = \frac{\partial}{\partial y}\left(k\frac{\partial T}{\partial y}\right)dV$$

$$\dot{q}_{Net,z} = \frac{\partial}{\partial z}\left(k\frac{\partial T}{\partial z}\right)dV$$

For steady heat flow, there is no net change in amount of energy stored in the soli, so the sum of the net rate of flow of energy in the three directions is zero. Thus,

$$\frac{\partial}{\partial x}\left(k\frac{\partial T}{\partial x}\right) + \frac{\partial}{\partial y}\left(k\frac{\partial T}{\partial y}\right) + \frac{\partial}{\partial z}\left(k\frac{\partial T}{\partial z}\right) = 0$$

Above Equation governs the steady diffusion of heat in a solid. when the thermal conductivity k is constant (i.e., neither a function of temperate or location), Above Equation simplifies to

$$T_{xx} + T_{yy} + T_{zz} = \nabla^2 T = 0$$

which is the Laplace equation.

For steady two-dimensional heat diffusion, Above Equation becomes

$$T_{xx} + T_{yy} = 0$$

In terms of the general second-order PDE defined by Eq. $Af_{xx} + Bf_{xy} + Cf_{yy} + Df_x + Ef_y + Ff = G$, $A = 1$, $B = 0$, and $C = 1$. The discriminant, $B^2 - 4AC$, is

$$B^2 - 4AC = 0^2 - 4(1)(1) = -4 < 0$$

Consequently, Equation $T_{xx} + T_{yy} = 0$ is an elliptic PDE.

The characteristics associated with Eq. $T_{xx} + T_{yy} = 0$ are determined by performing characteristic analysis.

In this case, becomes $\begin{bmatrix} A & B & C \\ dx & dy & 0 \\ 0 & dx & dy \end{bmatrix}\begin{bmatrix} f_{xx} \\ f_{xy} \\ f_{yy} \end{bmatrix} = \begin{bmatrix} -Df_x - Ef_y - F + G \\ d(f_x) \\ d(f_y) \end{bmatrix}$

$$\begin{bmatrix} 1 & 0 & 1 \\ dx & dy & 0 \\ 0 & dx & dy \end{bmatrix} \begin{bmatrix} T_{xx} \\ T_{xy} \\ T_{yy} \end{bmatrix} = \begin{bmatrix} 0 \\ d(T_x) \\ d(T_y) \end{bmatrix}$$

The characteristic equation corresponding to Eq. $T_{xx} + T_{yy} = 0$ is determined by setting the deter-

minant of the coefficient matrix of above Equation $\begin{bmatrix} 1 & 0 & 1 \\ dx & dy & 0 \\ 0 & dx & dy \end{bmatrix} \begin{bmatrix} T_{xx} \\ T_{xy} \\ T_{yy} \end{bmatrix} = \begin{bmatrix} 0 \\ d(T_x) \\ d(T_y) \end{bmatrix}$ equal to zero and

solving the resulting equation for the slopes of the characteristic paths. Thus,

$$(1)(dy)^2 + (1)(dx)^2 = 0$$

$$\frac{dy}{dx} = \pm\sqrt{-1}$$

Equation $\frac{dy}{dx} = \pm\sqrt{-1}$ shows that there are no real characteristics associated with the steady two-dimensional heat conduction equation. Physically, this implies that there are no preferred paths of information propagation, and that the domain of dependence and range of influence of every point is the entire solution domain. The temperature at every point depends on the temperature at all the other points, including the boundaries of the solution domain, and the temperature at each point influences the temperature at all the other points. The temperature distribution is continuous throughout the solution domain because there are no paths along which the derivative of temperature may be discontinuous.

Fundamental Solution

Fundamental solution of a linear partial differential equation is a solution of a partial differential equation $Lu(x) = 0$, $x \in \mathbf{R}^n$, with coefficients of class C^∞, in the form of a function $I(x, y)$ that satisfies, for fixed $y \in \mathbf{R}^n$, the equation

$$LI(x, y) = \delta(x - y), \qquad x \neq y,$$

Which is interpreted in the sense of the theory of generalized functions, Where δ is the delta-function. There is a fundamental solution for every partial differential equation with constant coefficients, and also for arbitrary elliptic equations. For example, for the elliptic equation

$$\sum_{i,j=1}^{n} \alpha_{ij} \frac{\partial^2 u}{\partial x_i \partial x_j} = 0$$

with constant coefficients α_{ij} forming a positive-definite matrix α, a fundamental solution is provided by the function

$$I(x,y) \begin{cases} \left[\left| \sum_{i,j=1}^{n} A_{ij} \left(x_i - y_i\right)\left(x_j - y_j\right) \right| \right]^{(2-n)/2}, & n > 2, \\ \log \left[\sum_{i,j=1}^{n} A_{ij} \left(x_i - y_i\right)\left(x_j - y_j\right) \right], & n = 2, \end{cases}$$

Where A_{ij} is the cofactor of a_{ij} in the matrix α.

Fundamental solutions are widely used in the study of boundary value problems for elliptic equations.

Green's Function

Green's functions are a device used to solve difficult ordinary and partial differential equations which may be unsolvable by other methods. The idea is to consider a differential equation such as

$$\frac{d^2 f(x)}{dx^2} + x^2\, f(x) = 0 \Rightarrow \left(\frac{d^2}{dx^2} + x^2 \right) f(x) = 0 \Rightarrow \mathcal{L} f(x) = 0.$$

Above, the notation $\mathcal{L} = \dfrac{d^2}{dx^2} + x^2$ is defined so that $\mathcal{L}$ s a differential operator; a linear combination of derivative operators times functions. As above, the differential equation can be represented by such an operator acting on a function. The Green's function in this case is the analogue of the inverse of $\mathcal{L}$:

$$G(x, y) \sim \mathcal{L}^{-1} \sim \left(\frac{d^2}{dx^2} + x^2 \right)^{-1}.$$

The idea is that the Green's function inverts the operator, so the inhomogeneous version of the above, $\mathcal{L} f(x) = g(x)$ can be solved by the analogue of $f(x) = G(x, y) g(x)$. The above correspondence in this case gives $\mathcal{L} f(x) \sim \mathcal{L} G(x, y) g(x) \sim \mathcal{L}\mathcal{L}^{-1} g(x) = g(x)$. The formal mathematics underlying this idea and why $G(x, y)$ is a function of two variables.

The inverse of a derivative added to functions and so on is not a very well-defined object; rigorous mathematics is required to derive and justify a more precise construction. As a result, constructing and solving for Green's functions is a delicate and difficult procedure in general.

Green's functions are widely used in electrodynamics and quantum field theory, Where the relevant differential operators are often difficult or impossible to solve exactly but can be solved perturbatively using Green's functions. In field theory contexts the Green's function is often called the propagator or two-point correlation function since it is related to the probability of measuring a field at one point given that it is sourced at a different point.

Definition of the Green's Function

Formally, a Green's function is the inverse of an arbitrary linear differential operator $\mathcal{L}$. It is a function of two variables $G(x,y)$ which satisfies the equation

$$\mathcal{L}G(x,y)=\delta(x-y)$$

With $\delta(x-y)$ the Dirac delta function. This says that the Green's function is the solution to the differential equation with a forcing term given by a point source. Informally, the solution to the same differential equation with an arbitrary forcing term can be built up point by point by integrating the Green's function against the forcing term. This is equivalent to taking an uncountable superposition of solutions to the equation with point source and adding them up to the arbitrary forcing term, which is why the linearity of the differential operator is important. Formally, this means the solution to an arbitrary linear differential equation with forcing term $\mathcal{L}u(x)=f(x)$ is given by

$$u(x)=\int G(x,y)f(y)\,dy,$$

since then

$$\mathcal{L}u(x)=\int \mathcal{L}G(x,y)f(y)\,dy,\ =\int \delta(x-y)f(y)\,dy=f(x).$$

In general, Green's functions are not in fact functions but rather distributions, which means they can be integrated against functions. Although the resulting integrals may be difficult or impossible to compute, they provide an immediate solution to arbitrary linear differential equations when possibly no solution may be found by other methods, which can at the very least be computed numerically.

Below, several methods for constructing Green's functions are outlined. Which method is optimal is highly context-dependent.

Method of Direct Integration

As given above, the solution to an arbitrary linear differential equation can be written in terms of the Green's function via

$$u(x)=\int G(x,y)f(y)\,dy.$$

Since the Green's function solves

$$\mathcal{L}G(x,y)=\delta(x-y)$$

and the delta function vanishes outside the point $x=y$, one method of constructing Green's functions is to instead solve the homogeneous linear differential equation $\mathcal{L}G(x)=0$ and impose the correct boundary conditions at $x=y$ to account for a delta function.

This process can be written more formally as follows:

Below, the discussion is restricted to the special case of monic (leading coefficient unity) second-order linear differential operators for simplicity. First, write down the general form of the solutions on either side of $x = y$:

$$G(x, y) = \begin{cases} c_1 G_1(x) + c_2 G_2(x) & x < y \\ d_1 G_1(x) + d_2 G_2(x) & x > y, \end{cases}$$

Where c_1, c_2, d_1, d_2 are constants and G_1, G_2 re the two homogeneous solutions to the differential equation.

Next, impose the two boundary conditions. This fixes two of the constants c_1, c_2, d_1, d_2 in terms of the other two.

Third, impose continuity of $G(x, y)$ at $x = y$. This fixes one of the two remaining constants.

Lastly, require that $\dfrac{dG}{dx}$ increase by one at the delta function. This comes from integration of the original differential equation around a small window on either side of y:

$$\Delta \frac{dG}{dx}\bigg|_{x=y} = \int_{y-\epsilon}^{y+\epsilon} \frac{d^2 G}{dx^2} \, dx = \int_{y-\epsilon}^{y+\epsilon} \delta(x - y) dx = 1.$$

The reason why $\dfrac{d^2}{dx^2}$ alone is considered out of all the possible terms in $\mathcal{L}$ is because solutions must be continuous at y; any other terms in the differential operator do not change on either side of y when integrated.

This condition on the change in the derivative fixes the last constant and therefore solves for the Green's function.

Example: Consider electromagnetic waves polarized in the z-direction propagating in one dimension in a lasing cavity of length L. He waves propagate in an active laser medium with current density $J(x)$. Maxwell's equation for the z-component of the electric field of these waves then reads

$$\left(\frac{d^2}{dx^2} - k^2 \right) E_z(x) = J(x)$$

Where the constant k is given by $k^2 = \dfrac{g\omega^2}{c^2}$ with c the speed of light, ω the angular frequency of the light, and g a constant called the gain coefficient.

If the cavity is walled by conducting mirrors, the boundary conditions are Dirichlet:

$$E_z(0) = E_z(L) = 0.$$

Find the general solution for the electric field between the mirrors.

The homogeneous equation is

$$\left(\frac{d^2}{dx^2} - k^2\right) G(x, y) = 0,$$

which has solutions given by exponentially growing and decaying exponentials:

$$G(x, y) = \begin{cases} c_1 e^{kx} + c_2 e^{-kx} & x < y \\ d_1 e^{kx} + d_2 e^{-kx} & x > y. \end{cases}$$

The boundary conditions $G(0, y) = G(L, y) = 0$ impose the constraints on the coefficients:

$$c_1 + c_2 = 0$$

$$d_1 e^{kL} + d_2 e^{-kL} = 0.$$

Solving for c_2 and d_2 and plugging into the expression for $G(x, y)$ yields

$$G(x, y) \begin{cases} 2c_1 \sinh(kx) & x < y \\ 2d_1 e^{-kx} \sinh(k(x - L)) & x > y. \end{cases}$$

Enforcing continuity at $x = y$ gives

$$2c_1 \sinh(ky) = 2d_1 e^{-ky} \sinh(k(y - L)),$$

and requiring the derivative to jump by unity yields

$$d_1 = \frac{(\coth(kL) + 1) \sinh(ky)}{2k}, \quad c_1 = \frac{\sinh(k(y - L))}{2k \sinh(kL)}.$$

Plugging all constants into $G(x, y)$ gives the Green's function:

$$G(x, y) = \begin{cases} \dfrac{\sinh(kx) \sinh(k(y - L))}{k \sinh(kL)} & x < y \\[4mm] \dfrac{\sinh(ky) \sinh(k(x - L))}{k \sinh(kL)} & x > y. \end{cases}$$

The general solution for the electric field given an arbitrary current profile $J(x)$ is therefore

$$E_z(x) = \int_0^L G(x, y) J(y) dy$$

$$= \int_0^x \frac{\sinh(ky) \sinh(k(x - L))}{k \sinh(kL)} J(y) dy + \int_x^L \frac{\sinh(kx) \sinh(k(x - L))}{k \sinh(kL)} J(y) dy.$$

Method of Eigenvector Expansion

If one knows the spectrum of a differential operator, the Green's function may be easily computed via the formula

$$G(x, y) = \sum_n \frac{1}{\lambda_n} u_n(x) u_n^*(y),$$

Where the λ_n are the eigenvalues corresponding to the normalized eigenfunctions u_n and the star denotes complex conjugate. This formula holds if the differential operator is a second-order differential operator of a special class called Sturm-Liouville operators in which all coefficient functions are continuous and the coefficient of the first-order term is differentiable (in general this condition can be extended to operators that are higher than second order, but these are not often physically motivated).

The motivation for this definition comes from thinking about solutions to the differential equation as being expanded in a basis of the eigenfunctions of the differential operator. It is straightforward to check that the above definition satisfies the criteria for a Green's function: consider the differential equation $\mathcal{L}u(x) = f(x)$, then

$$\mathcal{L}u(x) = \int \mathcal{L}G(x, y) f(y) \, dy$$

$$= \int \mathcal{L} \sum_n \frac{1}{\lambda_n} u_n(x) u_n^*(y) f(y) \, dy$$

$$= \sum_n \frac{1}{\lambda_n} \mathcal{L}u_n(x) \int u_n^*(y) f(y) \, dy$$

$$= \sum_n \frac{1}{\lambda_n} \lambda_n u_n(x) \int u_n^*(y) f(y) \, dy.$$

Above, $\mathcal{L}u_n(x)$ is replaced by $\lambda_n u_n(x)$ since the u_n are the eigenfunctions of $\mathcal{L}$. But now

$$\mathcal{L}u(x) = \sum_n u_n(x) \int u_n^*(y) f(y) \, dy = \sum_n u_n(x) \langle u_n \mid f \rangle = f(x).$$

In the second equality we have emphasized that the integral is just the usual inner product of functions, so that the right-hand side above is really just $f(x)$ expanded in a basis of eigenfunctions, so $\mathcal{L}u(x) = f(x)$ as expected. Therefore, this expression for the Green's function solves the given differential equation.

Example: In quantum mechanics, the equation of motion for the wave function of the quantum harmonic oscillator is the boundary-value eigenproblem in one dimension:

$$\left(\frac{d^2}{dx^2} - x^2 \right) \psi(x) = -E\psi(x)$$

with boundary conditions $\psi(\pm\infty) = 0$ and E the energy of the particle in some system of appropriate units so that all relevant coefficients are unity. The allowed values of the energy are $E = 2n + 1$ for $n \in \{0, 1, 2, ...\}$; the corresponding orthonormal eigenvectors are

$$\psi_n(x) = \frac{1}{\sqrt{2^n n!}} \pi^{-1/4} e^{-x^2/2} H_n(x),$$

with $H_n(x)$ the Hermite polynomials given by

$$H_n(x) = (-1)^n e^{x^2} \frac{d^n}{dx^n}\left(e^{-x^2}\right).$$

Find the exact solution to the quantum harmonic oscillator with the forcing term $(2x^2 + 4x + 8)e^{-x^2/2}$ and boundary conditions $\psi(\pm\infty) = 0$, that is, solve

$$\left(\frac{d^2}{dx^2} - x^2\right)\psi(x) = (2x^2 + 4x + 8)e^{-x^2/2}$$

For $\psi(x)$ subject to these boundary conditions.

Write down the forcing term as a sum of eigenfunctions of $\left(\dfrac{d^2}{dx^2} - x^2\right)$. This requires the lowest three eigenfunctions:

$$\psi_0(x) = \pi^{-1/4} e^{-x^2/2}$$

$$\psi_1(x) = \left(\frac{\pi}{4}\right)^{-1/4} 2xe^{-x^2/2}$$

$$\psi_2(x) = (4\pi)^{-1/4}\left(4x^2 - 2\right)e^{-x^2/2}.$$

Therefore the forcing term can be written as

$$(2x^2 + 4x + 8)e^{-x^2/2} = \frac{1}{2}(4\pi)^{1/4}\psi_2(x) + 2\left(\frac{\pi}{4}\right)^{1/4}\psi_1(x) + 9\pi^{1/4}\psi_0(x).$$

Now consider the decomposition of the Green's function for the quantum harmonic oscillator as a sum over eigenvectors:

$$G(x, y) = \sum_n \frac{-1}{2n+1}\psi_n(x)\psi_n^*(y)$$

$$= -\psi_0(x)\psi_0(y) - \frac{1}{3}\psi_1(x)\psi_1(y) - \frac{1}{5}\psi_2(x)\psi_2(y) + \dots.$$

Further terms are omitted because they will not be relevant: the eigenfunctions of the differential operator are orthonormal. Writing down the general solution in terms of the Green's function, one finds

$$\psi(x) = \int G(x, y) f(y)\,dy$$

$$= -\psi_0(x)\int \psi_0(y)\left(9\pi^{1/4}\psi_0(x)\right)dy$$

$$- \frac{1}{3}\psi_1(x)\int \psi_1(y)\left[2\left(\frac{\pi}{4}\right)^{1/4}\psi_1(x)\right]$$

$$-\frac{1}{5}\psi_2(x)\int \psi_2(y)\left[\frac{1}{2}(4\pi)^{1/4}\psi_2(x)\right]$$

The orthogonally of the eigenfunctions ensures that all other integrals in the expansion vanish. The normalization of the eigenfunctions gives the final result:

$$\psi(x) = -9\pi^{1/4}\psi_0(x) - \frac{2}{3}\left(\frac{\pi}{4}\right)^{1/4}\psi_1(x) - \frac{1}{10}(4\pi)^{1/4}\psi_2(x)$$

$$= \left(-9 - \frac{4}{3}x - \frac{1}{5}(2x^2 - 1)\right)e^{-x^2/2}.$$

The solution satisfies the given boundary conditions, as it must do because all of the eigenfunctions satisfy the boundary conditions as well.

Euler–Tricomi Equation

The Tricomi equation is a second-order partial differential equation of mixed elliptic-hyperbolic type for $u(x, y)$ with the form:

$$u_{xx} + xu_{yy} = 0.$$

It was first analyzed in the work by Francesco Giacomo Tricomi on the well-posedness of a boundary value problem. The equation is hyperbolic in the half plane $x < 0$, elliptic in the half plane $x > 0$, and degenerates on the line $x = 0$. Its characteristic equation is

$$dy^2 + xdx^2 = 0,$$

Whose solutions are

$$y \pm \frac{2}{3}(-x)^{\frac{3}{2}} = C$$

For any constant C, which are real for $x < 0$. The characteristics comprise two families of semicubical parabolas lying in the half plane $x < 0$, with cusps on the line $x = 0$. This is of hyperbolic degeneracy, for which the two characteristic families coincide, perpendicularly to the line $x = 0$.

References

- Cole, K.D.; Beck, J.V.; Haji-Sheikh, A.; Litkouhi, B. (2011), Heat Conduction Using Green's Functions (2nd ed.), CRC Press, ISBN 978-1-43-981354-6

- Partial-differential-equation: scholarpedia.org, Retrieved 21 May 2018

- Perona, P; Malik, J. (1990), "Scale-Space and Edge Detection Using Anisotropic Diffusion", IEEE Transactions on Pattern Analysis and Machine Intelligence, 12 (7): 629–639, doi:10.1109/34.56205

- Non-linear-partial-differential-equation: encyclopediaofmath.org, Retrieved 11 July 2018

- Fundamental-solution: encyclopediaofmath.org, Retrieved 19 April 2018

- Thambynayagam, R. K. M. (2011), The Diffusion Handbook: Applied Solutions for Engineers, McGraw-Hill Professional, ISBN 978-0-07-175184-1

- Hyperbolic-Partial-Differential-Equation: mathworld.wolfram.com, Retrieved 31 March 2018

- Unsworth, J.; Duarte, F. J. (1979), "Heat diffusion in a solid sphere and Fourier Theory", Am. J. Phys., 47 (11): 891–893, Bibcode:1979AmJPh..47..981U, doi:10.1119/1.11601

- Elliptic-equation, science: britannica.com, Retrieved 29 June 2018

- Greens-functions-in-physics: brilliant.org, Retrieved 21 May 2018

4

Boundary Value Problems and Conditions

A differential equation that is subject to a set of constraints is a boundary value problem. The solution to such a differential equation satisfies the boundary conditions. In order to completely understand boundary value problems and boundary conditions, it is vital to understand the fundamental aspects of initial value problem, Sturm–Liouville problem, Dirichlet problem and elliptic boundary value problem.

A Boundary value problem is a system of ordinary differential equations with solution and derivative values specified at more than one point. Most commonly, the solution and derivatives are specified at just two points (the boundaries) defining a two-point boundary value problem.

A two-point boundary value problem (BVP) of total order n on a finite interval (a,b) may be written as an explicit first order system of ordinary differential equations (ODEs) with boundary values evaluated at two points as

$$y'(x) = f(x, y(x)), \quad x \in (a, b), \quad g(y(a), y(b)) = 0$$

Here, $y, f, g \in R^n$ and the system is called explicit because the derivative y' appears explicitly. The n boundary conditions defined by g must be independent; that is, they cannot be expressed in terms of each other (if g is linear the boundary conditions must be linearly independent).

In practice, most BVPs do not arise directly in the form $y'(x) = f(x, y(x))$, $x \in (a, b)$, $g(y(a), y(b)) = 0$ but instead as a combination of equations defining various orders of derivatives of the variables which sum to n. In an explicit BVP system, the boundary conditions and the right hand sides of the ordinary differential equations (ODEs) can involve the derivatives of each solution variable up to an order one less than the highest derivative of that variable appearing on the left hand side of the ODE defining the variable. To write a general system of ODEs of different orders in the form $y'(x) = f(x, y(x))$, $x \in (a, b)$, $g(y(a), y(b)) = 0$, we can define y as a vector made up of all the solution variables and their derivatives up to one less than the highest derivative of each variable, then add trivial ODEs to define these derivatives. Such rewritten systems may not be unique and do not necessarily provide the most efficient approach for computational solution.

The words two-point refer to the fact that the boundary condition function g is evaluated at the solution at the two interval endpoints a and b unlike for initial value problems (IVPs) where the n initial conditions are all evaluated at a single point. Occasionally, problems arise where the function g is also evaluated at the solution at other points in (a,b). In these cases, we have a multipoint BVP. As shown in Ascher, a multipoint problem may be converted to a two-point problem by defining separate sets of variables for each subinterval between the points and adding boundary conditions which ensure continuity of the variables across the whole interval. Like rewriting the original BVP in the compact form $y'(x) = f(x, y(x))$, $x \in (a, b)$, $g(y(a), y(b)) = 0$, rewriting a multipoint problem as a two-point problem may not lead to a problem with the most efficient computational solution.

Most practically arising two-point BVPs have separated boundary conditions where the function g may be split into two parts (one for each endpoint):

$$g_a(y(a)) = 0, \qquad g_b(y(b)) = 0.$$

Here, $g_a \in R^s$ and $g_b \in R^{n-s}$ for some value s with $1 < s < n$ and where each of the vector functions g_a and g_b are independent. However, there are well-known, commonly arising, boundary conditions which are not separated; for example, consider periodic boundary conditions which, for a problem written in the form of equation $y'(x) = f(x, y(x))$, $x \in (a, b)$, $g(y(a), y(b)) = 0$, are

$$y(a) - y(b) = 0.$$

Existence and Uniqueness

Questions of existence and uniqueness for BVPs are much more difficult than for IVPs. Indeed, there is no general theory. Consider the IVP

$$y'(x) = f(x, y(x)), \quad y(a) = s$$

Corresponding to the ODE in $y'(x) = f(x, y(x))$, $x \in (a, b)$, $g(y(a), y(b)) = 0$. If this IVP has a solution for all choices of initial vectors s then the existence of a solution to $y'(x) = f(x, y(x))$, $x \in (a, b)$, $g(y(a), y(b)) = 0$ hinges on the solvability of the nonlinear system of equations

$$g(s, y(b; s)) = 0$$

Where $y(b; s)$ is the solution of the IVP $y'(x) = f(x, y(x))$, $y(a) = s$ evaluated at $x = b$ for the initial value $y(a) = s$. If there is a solution then it is the unique solution (among solutions of this type) if the nonlinear system $g(s, y(b; s)) = 0$ has just one solution s.

For linear BVPs, where the ODEs and boundary conditions are both linear, the equation $g(s, y(b; s)) = 0$ is a linear system of algebraic equations. Hence, generally there will be none, one or an infinite number of solutions, analogously to the situation with systems of linear algebraic equations.

In addition to the possibilities for linear problems, nonlinear problems can also have a finite number of solutions. Consider the following simple model of the motion of a projectile with air resistance:

$$y' = \tan(\phi),$$

$$v' = -\frac{g}{v}\tan(\phi) - vv\sec(\phi),$$

$$\phi' = -\frac{g}{v^2}.$$

These equations may be viewed as describing the planar motion of a projectile fired from a cannon. Here, y is the height of the projectile above the level of the cannon, v is the velocity of the projectile, and ϕ is the angle of the trajectory of the projectile with the horizontal. The independent variable x measures the horizontal distance from the cannon. The constant v represents air resistance (friction) and g is the appropriately scaled gravitational constant. This model neglects three–dimensional effects such as cross winds and the rotation of the projectile. The initial height is $y(0)=0$ and the muzzle velocity $v(0)$ for the cannon is fixed. The standard projectile problem is to choose the initial angle of the cannon and hence of the projectile, $\phi(0)$, so that the projectile will hit a target at the same height as the cannon at a distance $x=x_{end}$; that is, we require $y(x_{end})=0$. Altogether the boundary conditions are

$$y(0)=y(x_{end})=0, \quad v(0)$$

Does this BVP have a solution? Physical intuition suggests that it certainly does not for x_{end} beyond the range of the cannon for the fixed muzzle velocity $v(0)$. On the other hand, if x_{end} is small enough, we do expect a solution, but is there only one? To see that there is not, consider the case when the target is very close to the cannon. We can hit the target by shooting with an almost flat trajectory or by shooting high and dropping the projectile mortar-like on the target. That is, there are (at least) two solutions that correspond to initial angles $\phi(0)=\phi_{low}>0$ and $\phi(0)=\phi_{high}<\pi/2$. It turns out that there are exactly two solutions; see for an example.

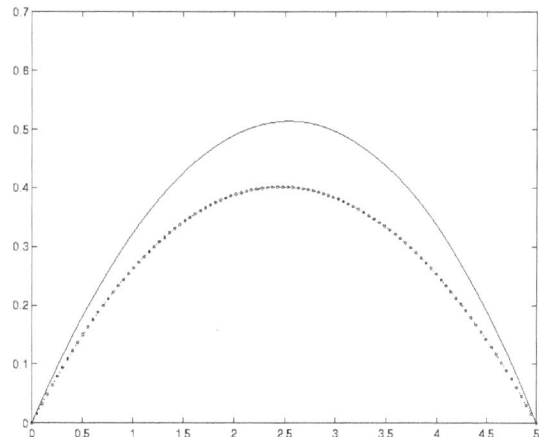

Figure : Two ways to hit a target at $x_{end}=5$ with initial velocity $v(0)=0.5$, friction $v=0.02$ and gravitational constant $g=0.032$

Now, let x_{end} increase. There are still two solutions, but the larger the value of x_{end} , the smaller the angle ϕ_{high} and the larger the angle ϕ_{low}. If we keep increasing x_{end} , eventually we reach the maximum range with the given muzzle velocity. At this distance there is just one solution;, that is, $\phi_{low}=\phi_{high}$. In summary, there is a critical value of x_{end} for which there is exactly one solution. If x_{end} is smaller than this critical value, there are exactly two solutions and if it is larger, there is no solution.

Shooting or Marching Methods

The approach to proving existence exemplified by the projectile model suggests a computational method of solution. This is to compute the unknown initial value $\phi(0)$ to satisfy the nonlinear equation $y(x_{end};\phi(0))=0$. This approach requires the (computational) solution of an IVP for the ODEs for each value of the angle $\phi(0)$ attempted. Also, the nonlinear equation may be solved by any suitable method. Since there are quality codes for both tasks this suggests an approach that can be useful in practice. Physical intuition suggests exploiting the relationship between the angle chosen and the range achieved in a bisection-like algorithm but, in more complex cases, such simple physical relationships are usually not available and a general purpose method such as a Newton iteration is often used. The shooting method can be very successful on simple problems such as the projectile problem. It can be extended easily to suggest a method of solution for almost any boundary value problem based on solving equation $g(s, y(b;s)) = 0$ and it has been automated in many pieces of mathematical software. However, its success depends on a number of factors the most important of which is the stability of the initial value problem that must be solved at each iteration. (An ODE problem is stable if a small change to the ODE and the the initial or boundary conditions leads to a small change in the solution.) Unfortunately, it is the case that for many stable boundary value problems the corresponding initial value problems (beginning from either endpoint and integrating towards the other endpoint) are insufficiently stable for shooting to succeed. So, shooting methods are not computationally suitable for the whole range of practical boundary value problems, particularly those on very long or infinite intervals. A second difficulty, sometimes interconnected with the aforementioned stability problem, is that methods such as Newton iteration for solving equation $g(s, y(b;s)) = 0$ may require a far more accurate initial estimate for the initial value s than is readily available.

Infinite Intervals

Many ODE BVPs arise from the analysis of partial differential equations through the computation of similarity solutions or via perturbation methods. These problems are often defined on semi-infinite ranges. For example,

$$f''' = \frac{f'' f}{2}, \quad f(0) = f'(0) = 0, \quad f'(\infty) = 1$$

The Blasius problem arises from a similarity solution of partial differential equations describing fluid flow over a flat plate. Of course, the boundary condition at infinity is asymptotic. It should be read as $f'(x) \to 1$ as $x \to \infty$ and it implies that $f(x) \sim x + C$ as $x \to \infty$ where the constant C is a priori unknown.

This problem is easy to solve computationally — shooting from the origin and using a standard nonlinear equation solver works without difficulty. Of course, we can't integrate the equations to infinity but we can replace the boundary condition at infinity by a corresponding one at a finite point, L, and that point L need not be chosen very large because the asymptotic expansion of the solution has $f(x) \sim x + C$ exponentially fast as $x \to \infty$. So, for example, using the boundary condition $f'(L) = 1$ with $L = 10$ provides a quite accurate solution. There are no fast increasing solutions to the equation near the desired solution so there is no unstable growth of computational solutions on quite long ranges of integration as long as the guess for the unknown initial value $f''(0)$ is not chosen too far away from the correct value.

In the Blasius problem, the location and type of boundary conditions are determined physically and give us a stable (well-conditioned) problem. In general, matters are more complicated though physical principles remain an essential guide. For simplicity of exposition (and understanding) consider the linear problem

$$y''' + 2y'' - y' - 2y = 0.$$

Its general solution is

$$y(x) = Ae^x + Be^{-x} + Ce^{-2x}$$

There are three components of the solution, two that decay as x increases from the origin towards infinity and one that grows. Suppose that we solve this equation on the interval $(0, \infty)$ with boundary conditions

$$y(0) = 1, \ y'(0) = 1, \ y(\infty) = 0.$$

The last boundary condition implies that $A = 0$. Then, the other boundary conditions imply that B=3 and C=−2. So, there is a unique solution of this BVP. On the other hand, if the boundary conditions are

$$y(0) = 1, \ y(\infty) = 0, \ y'(\infty) = 0,$$

The boundary condition $y(\infty) = 0$ again implies that $A = 0$, but now the third condition places no constraint on the coefficients, and the remaining condition tells us only that $C = 1 - B$, so any value of B results in a solution; that is, this BVP has infinitely many solutions. This problem provides an example of the requirements of exponential dichotomy. For a problem to be well-posed the boundary conditions must be set appropriately. For the simple equation $y''' + 2y'' - y' - 2y = 0.$, if the boundary conditions are separated, essentially we must have two boundary conditions at the origin and one at infinity matching the two decaying and one increasing (towards infinity) basis functions in the solution.

If a BVP with boundary conditions at infinity is not well-posed, it is natural to expect numerical difficulties when those boundary conditions are imposed at a large but finite point L even though, in this case, a solution may always be defined. Suppose then that we solve the equation $y''' + 2y'' - y' - 2y = 0.$ with boundary conditions

$$y(0) = 1, \ y(L) = 0, \ y'(L) = 0$$

replacing $y(0) = 1, \ y(\infty) = 0, \ y'(\infty) = 0$. For large values of L, the system of linear equations for the coefficients A , B , and C in the general solution is extremely ill–conditioned reflecting the poor stability (conditioning) of equation $y''' + 2y'' - y' - 2y = 0.$ with boundary conditions $y(0) = 1, \ y(\infty) = 0, \ y'(\infty) = 0.$

Numerical Methods

Various shooting methods have certain problems in their basic approaches. These problems may be overcome, at least partially, using variants on the shooting method which broadly come under the heading of multiple shooting.

Most general purpose software packages for BVPs are based on global methods which fall into two related categories. The first is finite differences where a mesh is defined on the interval (a, b) and the derivative in $y'(x) = f(x, y(x)), \quad x \in (a, b), \quad g(y(a), y(b)) = 0$ is replaced by a difference approximation at each mesh point. The resulting difference equations plus the boundary conditions give a set of algebraic equations for the solution on the mesh. These equations are generally non-linear but are linear when the differential equations and boundary conditions are both linear. To achieve a user-specified error the software generally adjusts the mesh placement using local error estimates based on higher order differencing involving techniques such as deferred correction.

A second global approach is to approximate the solution defined in terms of a basis for a linear space of functions usually defined piecewise on a mesh and to collocate this approximate solution. (In collocation we substitute the approximate solution in the system of ODEs then require the ODE system to be satisfied exactly at each collocation point. The number of collocation points plus the number of boundary conditions must equal the number of unknown coefficients in the approximate solution; that is, they must equal the dimension of the linear space.) The most common choice of approximation is a linear space of splines. For a given linear space, the collocation points must be placed judiciously to achieve optimal accuracy. The error is again controlled by adjusting the mesh spacing using local error estimates involving approximate solutions of varying orders of accuracy.

Choosing a spline basis for collocation (or more or less equivalently using certain types of Runge-Kutta formulas on the mesh) leads to a nonlinear system which must be solved iteratively. At each iteration we must solve a structured linear system of equations. When the boundary conditions are separated, the system is almost block diagonal. Similarly structured systems arise from finite difference approximations and also from multiple shooting techniques. Because of the great practical importance of this type of linear algebra problem, significant effort has been devoted to developing stable algorithms which minimize storage and maximize efficiency. The case of non separated boundary conditions leads to a similarly structured system whose solution poses potentially greater stability difficulties.

Sturm–Liouville Eigenproblems

Another type of BVP that arises in the analytical solution of certain linear partial differential equations is the Sturm–Liouville eigenproblem. In its simplest form this is a scalar self-adjoint linear second order ODE BVP

$$-(p(x)y'(x))' + q(x)y(x) = \lambda r(x)y(x), \qquad x \in (a,b), \qquad y(a) = y(b) = 0.$$

Here, the parameter λ, an eigenvalue, is to be determined such that the BVP $-(p(x)y'(x))' + q(x)y(x) = \lambda r(x)y(x)$, $x \in (a,b)$, $y(a) = y(b) = 0$ has a nontrivial (not identically zero) solution. There are broad analogies here with the generalized algebraic eigenproblem $Ax = \lambda Bx$ where, depending on the properties of the matrices A and B, various distributions of the finite number of eigenvalues λ are possible. In the case of the BVP $-(p(x)y'(x))' + q(x)y(x) = \lambda r(x)y(x)$, $x \in (a,b)$, $y(a) = y(b) = 0$, for simple cases there are a countable number of number of eigenvalues each with a corresponding solution $y(x)$ (an eigenfunction). So, for example, as shown in Zettl, if $p(x), q(x)$ and $r(x)$ are sufficiently smooth and $p(x), r(x) > 0$ on $[a,b]$ then the eigenvalues are real and distinct, and may be ordered $0 < \lambda_0 < \lambda_1 < \lambda_2 < \dots$ defining a discrete spectrum. The eigenfunction $y_n(x)$ corresponding to λ_n has n zeros in (a,b) and the set of eigenfunctions $\{y_i(x)\}_{i=0}^{\infty}$ is linearly independent. If we relax the smoothness conditions on the coefficients p, q and r , and permit these functions to take on a wider range of values, many different phenomena are observed from doubling of the eigenvalues to the occurrence of continuous spectra.

ODE eigenvalue problems can be solved using a general-purpose code shooting code that treats an eigenvalue as an unknown parameter. However, with such a code one can only hope to compute an eigenvalue close to a guess. Specialized codes are much more efficient and allow you to be sure of computing a specific eigenvalue for a survey. Numerical methods for Sturm–Liouville eigenproblems that have been implemented in software include finite difference and finite element discretizations which each lead to generalized algebraic eigenproblems where approximations to a number of the lower eigenvalues are available simultaneously. Other methods popularized by Pruess approximate the ODE eigenproblem by another where the coefficients p, q and r are replaced by piecewise constants; this results in a set of problems which may each be solved analytically, again producing approximations to a number of the lower eigenvalues. Finally, shooting methods are usually implemented using a scaled Prufer transformation,

$pu' = \sqrt{S}r\cos(\theta), \quad u = \dfrac{r\sin(\theta)}{\sqrt{S}}$ where S is a scaling function, $S = 1$ gives the standard Prufer trans-

formation. The transformation leads to a pair of nonlinear ODEs for r and θ where the ODE for θ does not depend on r so may be solved alone. More directly important, the boundary conditions in problem $-(p(x)y'(x))' + q(x)y(x) = \lambda r(x)y(x), \quad x \in (a,b), \quad y(a) = y(b) = 0.$ are replaced by $\theta(a, \lambda_k) = 0, \theta(b, \lambda_k) = k\pi$ which provide the basis for a shooting method where each eigenvalue may be determined by the solution of a single nonlinear algebraic equation.

Initial Value Problem

Most differential equations have more than one solution. For a first-order equation, the general solution usually involves an arbitrary constant C, with one particular solution corresponding to each value of C.

What this means is that knowing a differential equation that a function $y(x)$ satisfies is not enough information to determine $y(x)$. To find the formula for $y(x)$ precisely, we need one more piece of information, usually called an initial condition.

For example, suppose we know that a function $y(x)$ satisfies the differential equation

$y' = y.$

It follows that

$y(x) = Ce^x$

For some constant C. If we want to determine C, we need at least one more piece of information about the function $y(x)$. For example, if we also know that

$y(0) = 3,$

the value of C must be 3, and hence $y(x) = 3e^x$.

Initial Value Problems

An initial value problem consists of

- A first-order differential equation $y' = f(x, y)$, and

- An initial condition of the form $y(a) = b$.

For example,

$$y' = y, \quad y(0) = 3$$

is an initial value problem, whose solution is

$$y = 3e^x .$$

In general, we expect that every initial value problem has exactly one solution. We can find this solution using the following procedure.

Solving Initial Value Problems

Given an initial value problem

$$y' = f(x, y), \quad y(a) = b,$$

We can solve it using the following procedure:

- Find the general solution to the given differential equation, involving an arbitrary constant C.

- Substitute $x = a$ and $y = b$ to get an equation for C.

- Solve for C and then substitute the answer back into the formula for y.

Example:

Find the solution to the following initial value problem:

$$y' = -y^2 , \qquad\qquad y(0) = 5.$$

Solution:

The general solution of this differential equation:

$$y = \frac{1}{x + C} ,$$

Plugging in $x = 0$ and $y = 5$ gives the equation

$$5 = \frac{1}{0 + C} .$$

Solving for C gives $C = 1/5$, so

$$y = \frac{1}{x + (1/5)} \; .$$

This simplifies to

$$y = \frac{5}{5x + 1}$$

Example

Find the solution to the following initial value problem:

$$y' = 2y, \qquad y(0) = 5.$$

Solution: The given differential equation isn't very different from the equation

$$y' = y$$

In that case, the general solution was $y = Ce^x$. How can we modify this solution to account for the extra 2?

A few moments of thought reveals the answer:

$$y = Ce^{2x}$$

So this is the general solution to the given equation. Plugging in $x = 0$ and $y = 5$ gives the equation

$$5 = Ce^0$$

so $C = 5$ and the solution is

$$y = 5e^{2x}$$

Fundamental Theorem of ODE's (Optional)

As a general rule, we expect any initial value problem of the form

$$y' = f(x, y), \qquad y(a) = b$$

to have a unique solution. The following theorem gives specific conditions which guarantee that this holds.

Fundamental Theorem of ODE's

Consider an initial value problem of the form

$$y' = f(x, y), \qquad y(a) = b.$$

If the function $f(x, y)$ is continuously differentiable for all values of x and y, then this initial value problem has a unique solution.

Here continuously differentiable means that both partial derivatives

$$\frac{\partial f}{\partial x} \text{ and } \frac{\partial f}{\partial y}$$

exist and are continuous.

This theorem is also known as the existence and uniqueness theorem for firstorder ODE's, since it guarantees both that the solution exists and that it is unique.

The hypothesis that the function $f(x, y)$ is continuously differentiable is important for the theorem. In fact, there are initial value problems that do not satisfy this hypothesis that have more than one solution. For example, the initial value problem

$$y' = \frac{y}{x}, \qquad y(0) = 0$$

has infinitely many different solutions, namely the lines $y = Cx$ for all possible values of C. The function $f(x, y)$ in this case is y / x, which is not defined (and hence not continuously differentiable) when $x = 0$.

There is a nice geometric interpretation of the fundamental theorem. As we have seen, the solutions to a differential equation can be viewed as a family of solution curves in the xy -plane. For example, Below Figure shows the curves $y = \ln(x + C)$, which are the solutions to the differential equation

$$y' = e^{-y}.$$

From a geometric point of view, an initial condition $y(a) = b$ is the same as a point (a, b) that the solution curve must pass through. Thus, saying that the initial value problem

$$y' = f(x, y) \qquad y(a) = b$$

has a unique solution is the same as saying that the point (a, b) has exactly one solution curve passing through it. This leads us to the following restatement of the fundamental theorem of ODE's.

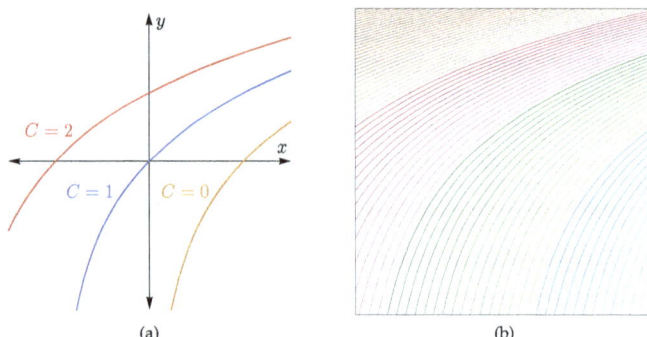

(a)   (b)

Figure: (a) Three curves of the form $y = \ln(x + C)$. (b) The family of all such curves completely fills the plane.

Fundamental Theorem of ODE's (Geometric Version)

Consider a first-order differential equation of the form

$$y' = f(x, y),$$

Where the function $f(x, y)$ is continuously differentiable. Then:

1. The solution curves for this differential equation completely fill the plane,

2. Solution curves for different solutions do not intersect.

Here statement (1) is the same as saying that every point (a, b) lies on at least one solution curve, i.e. every initial condition gives at least one solution. Statement (2) is the same as saying that no point (a, b) lies on more than one solution curve, i.e. every initial condition has at most one solution.

Example: Solve the initial value problem.

$$\frac{dy}{dx} = 10 - x \quad y(0) = -1$$

Solution:

Step 1:

$$\frac{dy}{dx} = 10 - x \quad \rightarrow \quad dy = (10 - x) \, dx$$

$$\int dy = \int (10 - x) \, dx \rightarrow y = 10x - \frac{x^2}{2} + c$$

Step 2: When $x = 0, y = -1$.

$$-1 = 10(0) - \frac{0}{2} + c \rightarrow \quad c = -1$$

$$\text{Solution}: y = 10x - \frac{x^2}{2} - 1$$

Example: Solve the initial value problem.

$$\frac{dy}{dx} = 9x^2 - 4x + 5, \quad y(-1) = 0$$

Solution:

Step 1:

$$\frac{dy}{dx} = 9x^2 - 4x + 5 \quad \rightarrow \quad dy = (9x^2 - 4x + 5) \, dx$$

$$\int dy = \int (9x^2 - 4x + 5) \, dx \quad \rightarrow \quad y = \frac{9x^3}{3} - \frac{4x^2}{2} + 5x + c$$

Step 2: When $x = -1$, $y = 0$.

$$0 = 3(-1)^3 - 2(-1)^2 + 5(-1) + c \quad \rightarrow \quad 0 = -3 - 2 - 5 + c \quad \rightarrow \quad c = 10$$

Solution: $\quad y = 3x^3 - 2x^2 + 5x + 10$

Example: Solve the initial value problem.

$$\frac{ds}{dt} = \cos t + \sin t, \qquad s(\pi) = 1$$

Solution:

Step 1:

$$\frac{ds}{dt} = \cos t + \sin t \quad \rightarrow \quad ds = (\cos t + \sin t)\, dt$$

$$\int ds = \int (\cos t + \sin t)\, dt \quad \rightarrow s = \sin t - \cos t + c$$

Step 2: When $t = \pi$, $s = 1$.

$$1 = \sin \pi - \cos \pi + c \quad \rightarrow \quad 1 = 0 - (-1) + c \quad \rightarrow \quad c = 0$$

Solution: $s = \sin t - \cos t$

Example: Solve the initial value problem.

$$\frac{d^2 y}{dx^2} = 2 - 6x, \quad y'(0) = 4, \quad y(0) = 1$$

Solution: We will have to do the two steps twice to find the solution to this initial value problem. The first time through will give us y' and the second time through will give us y.

Step 1:

$$y' = \int (2 - 6x)\, dx \rightarrow y' = 2x - \frac{6x^2}{2} + c$$

Step 2: When $x = 0$, $y' = 4$.

$$4 = 2(0) - 3(0)^2 + c \quad \rightarrow \quad c = 4$$

$$y' = 2x - 3x^2 + 4$$

Step 1:

$$y = \int (2x - 3x^2 + 4)\, dx \quad \rightarrow \quad y = \frac{2x^2}{2} - \frac{3x^3}{3} + 4x + c$$

Step 2: When $x = 0$, $y = 1$.

$$1 = 0^2 - 0^3 + 4(0) + c \quad \rightarrow \quad c = 1$$

Solution: $y = x^2 - x^3 + 4x + 1$

Example: Solve the initial value problem.

$$y^{(4)} = -\sin t + \cos t; \quad y'''(0) = 7, \quad y''(0) = y'(0) = -1,$$
$$y(0) = 0$$

Solution: Since we are working with the fourth derivative, we will have to go through the two steps four times.

Step 1:

$$y''' = \int (-\sin t + \cos t)\, dt \quad \rightarrow \quad y''' = \cos t + \sin t + c$$

Step 2: When $t = 0$, $y''' = 7$.

$$7 = \cos 0 + \sin 0 + c \quad \rightarrow \quad 7 = 1 + c \rightarrow c = 6$$
$$y''' = \cos t + \sin t + 6$$

Step 1:

$$y'' = \int (\cos t + \sin t + 6) dt \rightarrow y'' = \sin t - \cos t + 6t + c$$

Step 2: When $t = 0$, $y'' = -1$.

$$-1 = \sin 0 - \cos 0 + 6(0) + c \quad \rightarrow \quad -1 = -1 + c \rightarrow c = 0$$
$$y'' = \sin t - \cos t + 6t$$

Step 1:

$$y' = \int (\sin t - \cos t + 6t)\, dt \quad \rightarrow \quad y' = -\cos t - \sin t + \frac{6t^2}{2} + c$$

Step 2: When $t = 0$, $y' = -1$.

$$-1 = -\cos 0 - \sin 0 + 3(0)^2 + c \quad \rightarrow \quad -1 = -1 + c \rightarrow c = 0$$
$$y' = -\cos t - \sin t + 3t^2$$

Step 1:

$$y = \int (-\cos t - \sin t + 3t^2)\, dt \quad \rightarrow \quad y = -\sin t + \cos t + \frac{3t^3}{3} + c$$

Step 2: When t = 0, y = 0.

$$0 = -\sin 0 + \cos 0 + 0^3 + c \quad \rightarrow \quad 0 = 1 + c \quad \rightarrow \quad c = -1$$

Solution: $y = -\sin t + \cos t + t^3 - 1$

Example: Given the velocity,

$$v = \frac{ds}{dt} = 32t - 2,$$

and the initial position of the body as s $(1/2) = 4$. Find the body's position at time t.

Solution:

Step 1:

$$s = \int (32t - 2) \quad \rightarrow \quad s = \frac{32t^2}{2} - 2t + c$$

Step 2: When $t = 1/2$, $s = 4$.

$$4 = 16\left(\frac{1}{2}\right)^2 - 2\left(\frac{1}{2}\right) + c \quad \rightarrow \quad 4 = 4 - 1 + c \quad \rightarrow \quad c = 1$$

Solution: $s = 16t^2 - 2t + 1$

Example: Given the acceleration $a = d^2 s / dt^2 = -4\sin 2t$, initial velocity $v(0) = 2$, and the initial position of the body as $s(0) = -3$. Find the body's position at time t.

Solution:

Step 1:

$$v = \int -4 \sin 2t \, dt \quad \rightarrow \quad v = \frac{4\cos 2t}{2} + c$$

Step 2: when $t = 0$, $v = 2$.

$$2 = 2 \cos 0 + c \quad \rightarrow \quad 2 = 2 + c \quad \rightarrow \quad c = 0$$
$$v = 2 \cos 2t$$

Step 1:

$$s = \int 2 \cos 2t \, dt \quad \rightarrow \quad s = \frac{2 \sin 2t}{2} + c$$

Step 2: when $t = 0$, $s = -3$.

$$-3 = \sin 0 + c \quad \rightarrow \quad c = -3$$

Solution: $s = \sin 2t - 3$

Sturm–Liouville Problems

A classical "Sturm-Liouville equation", is a real second-order linear differential equation of the form

$$\frac{d}{dx}\left[p(x)\frac{dy}{dx}\right] + q(x)y = \lambda r(x)y,$$

In the simplest of cases all coefficients are continuous on the finite closed interval $[a, b]$, and $p(x)$ has continuous derivative. In this case y is called a"solution" if it is continuously differentiable on (a, b) and satisfies the equation $\frac{d}{dx}\left[p(x)\frac{dy}{dx}\right] + q(x)y = \lambda r(x)y$, at every point in (a, b). In addition, the unknown function y is required to satisfy boundary conditions. The function $r(x)$, is called the "weight" or "density" function.

The number of "famous" differential equations could be represented in the SL form:

- Bessel's equation

$$x^2 y" + xy' + (x^2 - v^2)y = 0$$

can be written in Sturm-Liouville form as

$$(xy')' + (x - v^2 / x)y = 0.$$

- The Legendre equation

$$(1 - x^2)y" - 2xy' + v(v+1)y = 0$$

can easily be put into SL form, since $(1 - x^2)' = -2x$, so, the Legendre equation is equivalent to

$$[(1 - x^2)y']' + v(v+1)y = 0$$

The general way to convert the 2nd order linear ODE to the SL form is to use an integrating factor $\mu(x)$ such that the equation

$$P(x)y" + Q(x)y' + R(x)y = 0$$

multiplied by $\mu(x)$ would have the SP form. One can easily show that

$$\mu(x) = \frac{1}{P(x)} \exp\left(\int Q(x)/P(x)dx\right)$$

does the job.

SL Boundary Value Problem (SL-BVP)

We introduce the SL-operator as

$$L[y] = \frac{d}{dx}\left[p(x)\frac{dy}{dx}\right] + q(x)y$$

and consider the SL equation

$$L[y] + \lambda r(x)y = 0 ,$$

Where $p > 0, r \geq 0$ and p, q and r are continuous functions on the interval $[a, b]$; along the with BC

$$B_a[y] = \alpha_1 y(a) + \alpha_2 p(a) y'(a) = 0 \quad and \quad B_b[y] = \beta_1 y(b) + \beta_2 p(b) y'(b) = 0 ,$$

Where $\alpha_1^2 + \alpha_2^2 \neq 0$ and $\beta_1^2 + \beta_2^2 \neq 0$

The problem of finding a complex number $\lambda = \mu$ such that the BVP $(L[y] + \lambda r(x)y = 0 ,)$, $(B_a[y] = \alpha_1 y(a) + \alpha_2 p(a) y'(a) = 0 \quad and \quad B_b[y] = \beta_1 y(b) + \beta_2 p(b) y'(b) = 0 ,)$ has a non-trivial solution is called SLP. The value $\lambda = \mu$ is called an eigenvalue and the corresponding solution $y(:, \mu)$ is called an eigenfunction.

There are three types of SLP:

1. A SLP is called regular $p > 0$, and $r > 0$ on (a, b)

2. A SLP is called singular if $p > 0$ on (a, b), $r \geq 0$ on (a, b) and $p(a) = p(b) = 0$

3. A SLP is called periodic if $p > 0$, $r > 0$ and p, q and r are continuous functions on (a, b); along with the following BC:

$$y(a) = y(b) \qquad\qquad y'(a) = y'(b) .$$

Example

For $\lambda \in R$ solve

$$y'' + \lambda y = 0, \quad y(0) = y'(\pi) = 0$$

We consider three cases corresponding to values of λ :

- $\lambda = -\mu^2 < 0$

The general solution of the ODE is given as

$$y = Ae^{-\mu x} + Be^{\mu x}$$

By substituting BC we obtain the following system:

$$\begin{cases} A + B & = 0 \\ -Ae^{-\mu\pi} + Be^{\mu\pi} & = 0 \end{cases}$$

This system has only trivial solution $A = B = 0$ (determinant of its matrix of coefficients is different from 0).

- $\lambda = 0$

In this case the problem has a solution $y = Ax + B$ and by substituting BC one can check that $A = B = 0$ (we get a trivial solution as well)

- $\lambda = \mu^2 > 0$

The general solution of the ODE is given as

$$y = A\,\cos(\mu x) + B\,\sin(\mu x)$$

By substituting BC we obtain the following system:

$$\begin{cases} A = 0 \\ B\cos(\mu\pi) = 0 \end{cases}$$

This problem has non-trivial solution (enabling $B \neq 0$) only when $\cos(\mu\pi) = 0$ or $\mu = (2n - 1)/2$. Therefore the eigenvalues λ_n could be written as

$$\lambda_n = \frac{(2n - 1)^2}{4} \ , \ n = 0, \pm 1, \pm 2, \ \ldots$$

and the eigenfunctions (we choose $B_n = 1$) are

$$\varphi_n = \sin\left(\sqrt{\lambda_n}\,x\right)$$

All the eigenvalues λ_n are positive and the eigenfunctions corresponding to each eigenvalue form a one dimensional vector space, and so the eigenfunctions are unique up to a constant multiple.

Example

For $\lambda \in R$ solve

$$y'' + \lambda y = 0, \ \ y(0) - y(\pi) = 0, \ y'(0) - y'(\pi) = 0$$

These BC are called "periodic BC".

We consider three cases corresponding to values of λ:

- $\lambda = -\mu^2 < 0$

The general solution of the ODE is given as

$$y = Ae^{-\mu x} + Be^{\mu x}$$

By substituting BC we obtain the following system:

$$\begin{cases} A(1 - e^{-\mu\pi}) + B(1 - e^{\mu\pi}) \ \ = 0 \\ A(-1 + e^{-\mu\pi}) + B(1 - e^{\mu\pi}) = 0 \end{cases}$$

This system has only trivial solution $A=B=0$ (for $\mu \neq 0$)

- $\lambda = 0$

In this case the problem has a solution $y = Ax + B$ and by substituting BC we obtain $A=0$ and B is an arbitrary constant. This corresponds to the eigenvalue $\lambda_0 = 0$ and the eigenfunction $\varphi_0 = 1$ (we set B = 1). Note that this eigenvalue is simple. The eigenvalue is called simple, if its eigenspace is of dimension one; otherwise the eigenvalue is called multiple.

- $\lambda = \mu^2 > 0$

The general solution of the ODE is given as

$$y = A \cos(\mu x) + B \sin(\mu x)$$

By substituting BC we obtain the following system:

$$\begin{cases} A(1 - \cos(\mu\pi)) - B \sin(\mu\pi) = 0 \\ A \sin(\mu\pi) + B(1 - \cos(\mu\pi)) = 0 \end{cases}$$

This problem has a non-trivial solution only when the determinant of the matrix of coefficients $D(\mu) = 2 - \cos(\mu\pi) = 0$. This corresponds to $\mu = 2n$, $n = \pm 1, \pm 2, \ldots$ and hence $\lambda_n = 4n^2$. The eigenfunctions corresponding to λ_n are given by (A = B = 1)

$$\varphi_n = \cos\left(\sqrt{\lambda_n}\, x\right) \qquad \psi_n = \sin\left(\sqrt{\lambda_n}\, x\right)$$

The all eigenvalues λ_n are positive and there are two linearly independent eigenfunctions corresponding to each eigenvalue, so they are not unique.

Properties of Sturm-Liouville System

It is interesting to note that a lot of information about the eigenvalues and eigenfunctions can be obtained without actually solving the SL problem. Some properties are that the eigenvalues are always real and bounded below but not above. If the interval (a, b) is finite, then eigenvalues are discrete. eigenfunctions are oscillatory in nature, and so on.

Sturm-Liouville Operator

Consider a regular SL problem

$$-\frac{d}{dx}\left[p(x)\frac{d}{dx}\right]y + q(x)y = \lambda\omega(x)y, \quad x \in [a, b]$$
$$c_a y(a) + d_a y'(a) = 0;$$
$$c_b y(b) + d_b y'(b) = 0.$$

Let $\mathcal{L}^2\left((a, b), \omega(x), dx\right)$ be the Hilbert space of square integrable functions on (a, b) with inner product

$$\langle f, g \rangle = \int_a^b \overline{f(x)} g(x) \, \omega(x) dx$$

with $\omega(x)$ called weight function. Let $\mathcal{H}$ be the subspace of functions that satisfy the boundary conditions of SL problem. Now, the differential operator of the form

$$L = \frac{1}{\omega(x)} \left[-\frac{d}{dx} \left[p(x) \frac{d}{dx} \right] + q(x) \right]$$

on some domain in $\mathcal{H}$, is called a Sturm-Liouville operator. Then the SL differential equation becomes an eigenvalue equation in the space $\mathcal{H}$.

$$Ly = \lambda y.$$

Theorem: Sturm-Liouville operator is self-adjoint operator on $\mathcal{H}$.

Proof:

$$\langle f, Lg \rangle = \int_a^b \overline{f(x)} (Lg)(x) \, \omega(x) dx$$

$$= \int_a^b \overline{f(x)} \left[-\frac{d}{dx} [p(x)g'(x)] + q(x)g(x) \right] dx$$

Integrating the first term by parts

$$\langle f, Lg \rangle = -\left[p(x)\overline{f(x)}g'(x) \right]_a^b + \int_a^b \left[\overline{f'(x)}p(x)g'(x) + \overline{f(x)}q(x)g(x) \right] dx$$

Similarly,

$$\langle f, Lg \rangle = -\left[p(x)\overline{f'(x)}g(x) \right]_a^b + \int_a^b \left[\overline{f'(x)}p(x)g'(x) + \overline{f(x)}q(x)g(x) \right] dx$$

Thus

$$\langle f, Lg \rangle - \langle Lf, g \rangle = -\left[p(x)\overline{f(x)}g'(x) \right]_a^b + \left[p(x)\overline{f'(x)}g(x) \right]_a^b$$

$$= p(b)\left(\overline{f'(b)}g(b) - \overline{f(b)}g'(b) \right) - p(a)\left(\overline{f'(a)}g(a) - \overline{f(a)}g'(a) \right)$$

Now, since both f, and g obey same boundary conditions,

$$c_a f(a) + d_a f'(a) = 0$$

$$\Rightarrow c_a \overline{f(a)} + d_a \overline{f'(a)} = 0$$

$$\text{and } c_a g(a) + d_a g'(a) = 0$$

$$\left(\overline{f'(a)}g(a) - \overline{f(a)}g'(a) \right) = 0$$

if $c_a \neq 0$ or $d_a \neq 0$. Simillary,

$$\left(\overline{f'(b)}g(b) - \overline{f(b)}g'(b) \right) = 0$$

Hence

$$\langle f, Lg \rangle = \langle Lf, g \rangle$$

Now, this result can easily be extended to periodic and singular SL systems. The two changes in the proof will be as follows:

- The subspace $\mathcal{H}$ will be appropriately defined by the BC.
- Note that rhs of the equation

$$\langle f, Lg \rangle - \langle Lf, g \rangle = - \left[p(x)\overline{f(x)}g'(x) \right]_a^b + \left[p(x)\overline{f'(x)}g(x) \right]_a^b$$
$$= p(b)\left(\overline{f'(b)}g(b) - \overline{f(b)}g'(b) \right) - p(a)\left(\overline{f'(a)}g(a) - \overline{f(a)}g'(a) \right)$$

will still be zero, if

  ○ $p(a) = p(b)$ and boundary conditions are periodic (equation $y(a) = y(b)$ and $y'(a) = y'(b)$) (That is if SL system is periodic);

  ○ $p(a) = 0$ and boundary condition at b is homogeneous (Singular SL system);

  ○ p(b) = o and boundary condition at a is homogeneous (Singular SL system);

  ○ Interval (a, b) is innite (since at innity functions will be vanishing) (Singular SL systems).

Regular SLP

Properties:

1. The eigenvalues of the regular SLP are real

    Proof: Suppose $\lambda \in C$ is an eigenvalue of the regular SLP and let y be corresponding eigenfunction. That is

$$L[y] + \lambda r(x)y = 0, \quad \alpha_1 y(a) + \alpha_2 p(a)y'(a) = 0, \quad \beta_1 y(b) + \beta_2 p(b)y'(b) = 0,$$

Taking the complex conjugates we get

$$L[\bar{y}] + \bar{\lambda} r(x)\bar{y} = 0, \quad \alpha_1 \bar{y}(a) + \alpha_2 p(a)\bar{y}'(a) = 0, \quad \beta_1 \bar{y}(b) + \beta_2 p(b)\bar{y}'(b) = 0,$$

Multiplying the ODE in

$$L[y] + \lambda r(x)y = 0, \quad \alpha_1 y(a) + \alpha_2 p(a)y'(a) = 0, \quad \beta_1 y(b) + \beta_2 p(b)y'(b) = 0, \text{ with } \bar{y}$$

and the ODE in

$$L[\bar{y}] + \bar{\lambda}r\,(x)\bar{y} \;=\; 0,\; \alpha_1\bar{y}(a)+\alpha_2 p(a)\bar{y}'\,(a)=0, \qquad \beta_1\bar{y}(b)+\beta_2 p(b)\bar{y}'(b) \;=\; 0\,,\text{with } y$$

and subtracting one from another yields

$$p(y'\bar{y} - y\bar{y}'\,)]' + \left(\lambda - \bar{\lambda}\right)r y\bar{y} \;=\; 0$$

Integrating the last expression we obtain

$$\left[p(y'\bar{y} - y\bar{y}')\right]\Big|_a^b = -\left(\lambda - \bar{\lambda}\right)\int_a^b r|y|^2\,dx = 0$$

The LHS of the last identity is zero due to the BC. Thus we have

$$\left(\lambda - \bar{\lambda}\right)\int_a^b r|y|^2\,dx = 0$$

From the definition of regular SLP we know that $r>0$ and y as an eigenfunction is different from zero as well. Therefore the only way to satisfy the identity is to set $\lambda = \bar{\lambda}$, which means that λ is real.

2. The eigenfunctions of a regular SLP corresponding to the distinct eigenvalues are orthogonal w.r.t. the weight function r(x) on (a, b). By other words, if the eigenfunctions u and v correspond to the distinct eigenvalues λ and μ then

$$\int_a^b r(x)u(x)v(x)dx \;=\; 0$$

Proof: As in the previous case we write the SL equations for functions u and v, multiply one for u by v and vice versa and subtract one equation from another. As result we get

$$\left[p(u'v - v'u)\right]' + \left(\lambda - \mu\right)ruv \;=\; 0$$

Integrating the last expression we obtain

$$\left[p(u'v - uv')\right]\Big|_a^b = -\left(\lambda - \mu\right)\int_a^b ruvdx = 0$$

The LHS of the last identity is zero due to the BC. Thus we have

$$\left(\lambda - \mu\right)\int_a^b ruvdx = 0$$

We know that $\lambda \neq \mu$ therefore $\int_a^b ruvdx = 0$ which confirms orthogonality.

3. The eigenvalues of the regular SLP are simple. Thus an eigenfunction that corresponds to an eigenvalue is unique up to a constant multiple.

4. The regular SL operator L is self-adjoint: if whenever $u,\,v \in C^2\,(a,\,b)\,\cap\,C^1[a,\,b]$ and satisfy the regular SLP and BC then

$$\int_a^b \left(vL[u] - uL[v]\right) dx = 0$$

Proof:

$$\int_a^b vL[u]\,dx = \int_a^b \left[-v(pu')\right] +vqu\ ]dx = (\text{integration by parts})$$

$$- pu'v\,|_a^b \ + \int_a^b (pu'v'+ quv)\,dx$$

By symmetry we see that

$$\int_a^b uL[v]\,dx = \int_a^b \left[-u(pv')\right] +uqv\ ]dx = (\text{integration by parts})$$

$$- puv'\,|_a^b \ + \int_a^b (pu'v'+ quv)\,dx$$

Subtracting one expression from another and applying BC $B_a[u] = B_a[v] = B_b[u] \ = \ B_b[v] = 0$ we obtain:

$$\int_a^b \left(vL[u]-uL[v]\right) dx = \left[p(u'v \ - \ uv')\right]|_a^b = 0.$$

Theorem: A self-adjoint regular SLP has an infinite number of real eigenvalues λ_n that are simple and satisfying

$$\lambda_1 < \lambda_2 < ... < \lambda_n < ...$$

With $\lim_{n\to\infty} \lambda_n = \infty$

Periodic SLP

Properties:

- The eigenvalues (if any) of the periodic SLP are real

- The eigenfunctions of a periodic SLP corresponding to the distinct eigenvalues are orthogonal w.r.t. the weight function r(x) on (a, b). By other words, if the eigenfunctions u and v correspond to the distinct eigenvalues λ and μ then

$$\int_a^b r(x)u(x)v(x)\,dx \ = \ 0$$

- The eigenvalues of the regular SLP are not simple. Thus an eigenfunction that corresponds to an eigenvalue is not unique.

- The periodic SL operator L is self-adjoint

All the proofs for periodic SLP are similar to the regular SLP and rely on the BC.

Theorem: A self-adjoint periodic SLP has an infinite number of real eigenvalues λ_n satisfying

$$-\infty < \lambda_1 < \lambda_2 \le \lambda_3 \le ... \le \lambda_n \le ...$$

The first eigenvalue λ_1 is simple. The number of linearly independent eigenfunctions correspond-ing to any eigenvalue $\lambda_n = \mu$ ($n > 1$) is equal to the repeated number of times μ.

Representation of Solutions and Numerical Calculation

The SL equation with boundary conditions may be solved in practice by a variety of numerical meth-ods. In difficult cases, one may need to carry out the intermediate calculations to several hundred decimal places of accuracy in order to obtain the eigenvalues correctly to a few decimal places.

There are the most common methods used for this purpose:

- Shooting methods: These methods proceed by guessing a value of λ, solving an initial value problem defined by the boundary conditions at one endpoint, say, a, of the interval $[a, b]$, comparing the value this solution takes at the other endpoint b with the other desired boundary condition, and finally increasing or decreasing λ as necessary to correct the orig-inal value. This strategy is not applicable for locating complex eigenvalues.

- Finite difference method: The method involves approximation of SL on the small sub-in-tervals of $[a, b]$ using the first few terms of Taylor series expansion of y.

- The Spectral Parameter Power Series (SPPS) method: It makes use of a generalization of the following fact about second order ordinary differential equations: if y is a solution which does not vanish at any point of (a, b), then the function

$$y(x) \int_a^x \frac{dt}{p(t)y(t)^2}$$

is a solution of the same equation and is linearly independent from y. Further, all solutions are linear combinations of these two solutions. In the SPPS algorithm, one must begin with an arbitrary value λ_0^* (often $\lambda_0^* = 0$; it does not need to be an eigenvalue) and any solution y_0 of SLP with $\lambda = \lambda_0^*$ which does not vanish on (a, b).

Two sequences of functions X^n and $\tilde{X}^n$ referred to as "iterated integrals", are defined re-cursively as follows. First when n = 0, they are taken to be identically equal to 1 on (a, b). To obtain the next functions they are multiplied alternately by $1/(py_0^2)$ and ry_0^2 and inte-grated, specifically

$$X^n(x) = -\int_a^x X^{n-1}(t) p(t)^{-1} y_0(t)^{-2} dt, \ n \text{ is odd}$$

$$X^n(x) = \int_a^x X^{n-1}(t) r(t) y_0(t)^2 dt, n \text{ is even}$$

$$\tilde{X}^n(x) = -\int_a^x \tilde{X}^{n-1}(t) p(t)^{-1} y_0(t)^{-2} dt, \ n \text{ is even}$$

$$\tilde{X}^n(x) = \int_a^x \tilde{X}^{n-1}(t) r(t) y_0(t)^2 dt, \ n \text{ is odd}$$

when n > 0. The resulting iterated integrals are now applied as coefficients in the following two power series in λ :

$$u_0 = y_0 \sum_{k=0}^{\infty} \left(\lambda - \lambda_0^*\right)^k \tilde{X}^{2k} \text{ and } u_1 = y_0 \sum_{k=0}^{\infty} \left(\lambda - \lambda_0^*\right)^k X^{2k+1}.$$

Then for any λ (real or complex), c_0 and c_1 are linearly independent solutions of the corresponding SL equation. (The functions p(x) and q(x) take part in this construction through their influence on the choice of y_0) Next one chooses coefficients c_0 and c_1 so the combination $y = c_0 u_0 + c_1 u_1$ satisfies the first boundary condition $B_a[y]$. This is simple to do since $X''(a) = \tilde{X}''(a) = 0$. The values of $X''(a)$ and $\tilde{X}''(b)$ provide the values of $u_0(b)$ and $u_1(b)$ and the derivatives $u_0'(b)$ and $u_1'(b)$, so the second boundary condition $B_b[y]$ becomes an equation in a power series in λ. For numerical work one may truncate this series to a finite number of terms, producing a calculable polynomial in λ whose roots are approximations of the sought-after eigenvalues. The SPPS method can, itself, be used to find a starting solution y_0.

Example: application of SLP to solution of PDEs

For a linear second order in one spatial dimension and first order in time of the form:

$$f(x)\frac{\partial^2 u}{\partial x^2} + g(x)\frac{\partial u}{\partial x} + h(x)u = \frac{\partial u}{\partial t} + k(t)u$$

$$u(a,\ t) = u(b,\ t) = 0$$

$$u(x,\ 0) = s(x)$$

Let us apply separation of variables, which in doing we must impose that:

$$u(x,\ t) = X(x)T(t)$$

Then above PDE may be written as:

$$\frac{\hat{L}X(x)}{X(x)} = \frac{\hat{M}T(t)}{T(t)}$$

Where

$$\hat{L} = f(x)\frac{d^2}{dx^2} + g(x)\frac{d}{dx} + h(x) \text{ and } \hat{M} = \frac{d}{dt} + k(t)$$

Since, by definition, $\hat{L}$ and $X(x)$ are independent of time t and $\hat{M}$ and $T(t)$ are independent of position x, then both sides of the above equation must be equal to a constant that we call λ. In such a case:

$$\hat{L}X(x) = \lambda X(x)$$

$$X(a) = X(b) = 0$$

$$\hat{M}T(t) = \lambda T(t).$$

The first of these equations must be solved as a SLP. Since there is no general analytic (exact) solution to SLP we can assume we already have the solution to this problem, that is, we have the eigenfunctions $X_n(x)$ and eigenvalues λ_n. The second of these equations can be analytically solved once the eigenvalues are known:

$$\frac{d}{dt}T_n(t) = \left(\lambda_n - k(t)\right)T_n(t)$$

$$T_n(t) = a_n e^{-\left(\lambda_n t - \int_0^t k(\tau)d\tau\right)}$$

and finally

$$u(x, t) = \sum_n a_n X_n(x) e^{-\left(\lambda_n t - \int_0^t k(\tau)d\tau\right)}$$

where

$$a_n = \frac{\left(X_n(x), s(x)\right)}{\left(X_n(x), X_n(x)\right)}$$

We compute a_n as an inner product defined as

$$\left(y(x), z(x)\right) = \int_a^b y(x)z(x)w(x)dx$$

where (for our case)

$$w(x) = \frac{e^{\int \frac{g(x)}{f(x)}dx}}{f(x)}$$

Dirichlet Problem

Dirichlet problem in mathematics is the problem of formulating and solving certain partial differential equations that arise in studies of the flow of heat, electricity, and fluids. Initially, the problem was to determine the equilibrium temperature distribution on a disk from measurements taken along the boundary. The temperature at points inside the disk must satisfy a partial differential equation called Laplace's equation corresponding to the physical condition that the total heat energy contained in the disk shall be a minimum. A slight variation of this problem occurs when there are points inside the disk at which heat is added (sources) or removed (sinks) as long as the temperature still remains constant at each point (stationary flow), in which case Poisson's equation is satisfied. The Dirichlet problem can also be solved for any simply connected region—i.e., one containing no holes—if the temperature varies continuously along the boundary. The problem is named for the 19th-century German mathematician Peter Gustav Lejeune Dirichlet, who suggested the first general method of solving this class of problems.

Dirichlet Problem for the Disk

The Dirichlet problem in a disk of radius r_0 and center at $(0, 0)$ can be expressed as

$$PDE: U_{rr} + \frac{U_r}{r} + \frac{U_{\theta\theta}}{r^2} = 0, \quad 0 < r < r_0, \quad -\pi \leq \theta \leq \pi,$$

$$BC: U(r_0, \theta) = f(\theta) \quad -\pi \leq \theta \leq \pi,$$

where $f(\theta)$ is a given periodic, continuous function of period 2π $(f(\theta + 2\pi) = f(\theta))$. To solve the above problem, we use the method of separation of variables.

Step 1. (Writing the ODEs): Seek solutions of the form

$$U(r, \theta) = R(r)T(\theta),$$

Where $0 \leq r \leq r_0$ and $-\pi \leq \theta \leq \pi$. Substituting into

$$PDE: U_{rr} + \frac{U_r}{r} + \frac{U_{\theta\theta}}{r^2} = 0, \quad 0 < r < r_0, \quad -\pi \leq \theta \leq \pi,$$

$$BC: U(r_0, \theta) = f(\theta) \quad -\pi \leq \theta \leq \pi,$$

and separating variables yield

$$R''(r)T(\theta) + r^{-1}R'(r)T(\theta) + r^{-2}R(r)T''(\theta) = 0.$$

$$\Rightarrow \frac{r^2R''(r) + rR'(r)}{R(r)} = -\frac{T''(\theta)}{T(\theta)} = k.$$

Which leads to the following two ODEs:

$$T''(\theta) + kT(\theta) = 0,$$

$$r^2 R''(r) + rR'(r) - kR(r) = 0.$$

Step 2. (Solving the ODEs):

Case (a): When $k < 0$, the general solution to $T''(\theta) + kT(\theta) = 0$, is the sum of two exponentials. Hence we have only trivial 2π -periodic solutions.

Case (b): When $k = 0$, we find that $T(\theta) = A\theta + B$ is the solution to $T''(\theta) + kT(\theta) = 0$. This linear function is periodic only when $A = 0$, that is, $T_0(\theta) = B$ is the only   -periodic solution corresponding to k = 0.

Case (c): When $k > 0$, the general solution to $T''(\theta) + kT(\theta) = 0$, is

$$T(\theta) = A \cos\left(\sqrt{k}\theta\right) + B \sin\left(\sqrt{k}\theta\right).$$

In this case we get a nontrivial 2π-periodic solution only when $\sqrt{k} = n$, $n = 1, 2, \ldots$. Hence, we obtain the nontrivial 2π-periodic solutions

$$T_n(\theta) = A_n \cos(n\theta) + B_n \sin(n\theta)$$

Corresponding to $\sqrt{k} = n$, $n = 1, 2, \ldots$.

Now for $k = n^2$, $n = 0, 1, 2, \ldots$, equation $r^2 R''(r) + r R'(r) - kR(r) = 0.$ is the Cauchy-Euler equation

$$r^2 R''(r) + r R'(r) - n^2 R(r) = 0.$$

When $n = 0$, the general solution is

$$R_0(r) = C + D \ln r.$$

Since $\ln r \to \infty$ as $r \to 0^+$, this solution is unbounded near $r = 0$ when $D \neq 0$. Therefore, we must choose $D = 0$ if $U(r, \theta)$ is to be continuous at $r = $ o. We now have $R_0(r) = C$ and so $U_0(r, \theta) = R_0(r) T_0(\theta) = CB$. For convenience, we write $U_0(r, \theta)$ in the form

$$U_0(r, \theta) = \frac{A_0}{2},$$

where A_0 is an arbitrary constant.

When $k = n^2$, $n = 1, 2, \ldots$, the general solution of $r^2 R''(r) + r R'(r) - kR(r) = 0.$ is given by

$$R_n(r) = C_n r^n + D_n r^{-n}.$$

Since $r^- \to \infty$ as $r \to 0^+$, we must set $D_n = 0$ in order for $u(r, \theta)$ to be bounded at $r = 0$. Thus

$$R_n(r) = C_n r^n$$

Now for each $n = 1, 2, \ldots$, we have the solutions

$$U(r, \theta) = R_n(r) T_n(\theta) = C_n r^n \left[A_n \cos(n\theta) + B_n \sin(n\theta) \right].$$

By superposition principle, we write

$$U(r, \theta) = \frac{A_0}{2} + \sum_{n=1}^{\infty} C_n r^n \left[A_n \cos(n\theta) + B_n \sin(n\theta) \right].$$

This series may be written in the equivalent form

$$U(r, \theta) = \frac{A_0}{2} + \sum_{n=1}^{\infty} \left(\frac{r}{r_0} \right)^n \left[A_n \cos(n\theta) + B_n \sin(n\theta) \right],$$

Where the A_n's and b_n's are constants. These constants can be determined from the boundary condition. With $r = r_0$ in $U(r, \theta) = \dfrac{A_0}{2} + \sum\limits_{n=1}^{\infty} \left(\dfrac{r}{r_0}\right)^n \left[A_n \cos(n\theta) + B_n \sin(n\theta)\right]$, we have

$$f(\theta) = \frac{A_0}{2} + \sum_{n=1}^{\infty} \left[A_n \cos(n\theta) + B_n \sin(n\theta)\right].$$

Since $f(\theta)$ is 2π-periodic, we recognize that A_n, B_n are Fourier coefficients. Thus

$$A_n = \frac{1}{\pi} \int_{-\pi}^{\pi} f(\theta) \cos(n\theta)\,d\theta, \ n = 0, 1, \ldots,$$

$$B_n = \frac{1}{\pi} \int_{-\pi}^{\pi} f(\theta) \sin(n\theta)\,d\theta, \ n = 1, \ldots,$$

We now summarize the Dirichlet problem for a disk as follows.

In the Dirichlet problem
$$PDE: U_{rr} + \frac{U_r}{r} + \frac{U_{\theta\theta}}{r^2} = 0, \quad 0 < r < r_0, \quad -\pi \le \theta \le \pi,$$
$$BC: U(r_0, \theta) = f(\theta) \quad -\pi \le \theta \le \pi,$$
, if

$$f(\theta) = \frac{A_0}{2} + \sum_{n=1}^{\infty} \left[A_n \cos(n\theta) + B_n \sin(n\theta)\right],$$

then the solution is given by

$$U(r, \theta) = \frac{A_0}{2} + \sum_{n=1}^{\infty} \left(\frac{r}{r_0}\right)^n \left[A_n \cos(n\theta) + B_n \sin(n\theta)\right],$$

where A_n and B_n are given by $A_n = \dfrac{1}{\pi} \int_{-\pi}^{\pi} f(\theta) \cos(n\theta)\,d\theta, \ n = 0, 1, \ldots,$ and

$B_n = \dfrac{1}{\pi} \int_{-\pi}^{\pi} f(\theta) \sin(n\theta)\,d\theta, \ n = 1, \ldots,$ respectively.

Example: Solve the following BVP

$$PDE: U_{rr} + \frac{U_r}{r} + \frac{U_{\theta\theta}}{r^2} = 0, \quad 0 < r < 1,$$
$$BC: U(1, \theta) = f(\theta)$$

Where $f(\theta) = 1 + r \sin\theta + \dfrac{r^3}{2} \sin(3\theta) + r^4 \cos(4\theta)$.

Solution: Here $r_0 = 1$. Note that f(θ) is already in the form of Fourier series, with

$$A_n = \begin{cases} 2 \text{ for n} = 0 \text{ and 1 for n} = 4 \\ 0 \text{ for other n} \end{cases} \qquad B_n = \begin{cases} 1 & n = 1 \\ \dfrac{1}{2} & n = 3 \\ 0 & \text{for other n} \end{cases}$$

The solution of the BVP is

$$U(r,\ \theta) = \frac{A_0}{2} + \sum_{n=1}^{\infty} \left(\frac{r}{r_0}\right)^n \left[A_n \cos(n\theta) + B_n \sin(n\theta)\right],$$

$$= 1 + r \sin\theta + \frac{r^3}{2} \sin(3\theta) + r^4 \cos(4\theta)$$

Exterior Dirichlet Problem: We shall discuss the exterior Dirichlet problem i.e., the Dirichlet problem outside the circle. The exterior Dirichlet problem is given by

$$PDE: U_{rr} + \frac{U_r}{r} + \frac{U_{\theta\theta}}{r^2} = 0, \quad 1 < r < \infty,$$

$$BC: U(1,\theta) = f(\theta), \quad 0 \le \theta \le 2\pi.$$

This problem is solved exactly in a manner similar to the interior Dirichlet problem. We assume that the solutions are bounded as $r \to \infty$. Basically, we throw out the solutions

$$r^n \cos(n\theta), \quad r^n \sin(n\theta), \quad \ln r$$

that are unbounded as $r \to \infty$.

The solution is given by

$$U(r,\ \theta) = \sum_{n=0}^{\infty} r^{-n} \left[A_n \cos(n\theta) + B_n \sin(n\theta)\right],$$

where A_n and B_n are given by

$$A_0 = \frac{1}{2\pi} \int_0^{2\pi} f(\theta) d\theta,$$

$$A_n = \frac{1}{\pi} \int_0^{2\pi} f(\theta) \cos(n\theta) d\theta,$$

$$B_n = \frac{1}{\pi} \int_0^{2\pi} f(\theta) \sin(n\theta) d\theta,$$

Boundary Conditions

Dirichlet Boundary Condition

The Dirichlet boundary condition is a type of boundary condition named after Peter Gustav Lejeune Dirichlet.

Peter Gustav Lejeune Dirichlet

This condition specifies the value that the unknown function needs to take on along the boundary of the domain. Given, for example, the Laplace equation, the boundary value problem with the Dirichlet b.c. is written as:

$$\Delta\varphi(\underline{x}) = 0 \qquad \forall \underline{x} \in \Omega$$

$$\varphi(\underline{x}) = f(\underline{x}) \qquad \forall \underline{x} \in \partial\Omega$$

Where φ is the unknown function, $\underline{x}$ is the independent variable (e.g. the spatial coordinates), Ω is the function domain, $\partial\Omega$ is the boundary of the domain, and f is a given scalar function defined on $\partial\Omega$. In the framework of numerical simulations, it is usually imposed directly in the algebraic system to be solved. Let's consider the following algebraic system derived from a numerical algorithm:

$$\begin{bmatrix} k_{1,1} & k_{1,2} & \cdot & k_{1,m-1} & k_{1,m} \\ k_{2,1} & k_{2,2} & \cdot & k_{2,m-1} & k_{2,m} \\ \cdot & \cdot & \cdot & & \cdot \\ k_{m-1,1} & k_{m-1,2} & \cdot & k_{m-1,m-1} & k_{m-1,m} \\ k_{m,1} & k_{m,2} & \cdot & k_{m,m-1} & k_{m,m} \end{bmatrix} \begin{bmatrix} x_1 \\ x_2 \\ \cdot \\ x_{n-1} \\ x_n \end{bmatrix} = \begin{bmatrix} a_1 \\ a_2 \\ \cdot \\ a_{m-1} \\ a_m \end{bmatrix}$$

Where k_{ij} are the elements of the algebraic operator (e.g. the stiffness matrix), x_i are the unknowns (i.e. the degrees of freedom of the problem), and a_i are the known terms. The simplest way to impose a Dirichlet boundary condition on the nth degree of freedom is to modify the system as follows:

$$\begin{bmatrix} k_{1,1} & \cdot & \cdot & \cdot & k_{1,m} \\ \cdot & \cdot & \cdot & \cdot & \cdot \\ 0 & 0 & 1 & 0 & 0 \\ \cdot & \cdot & \cdot & \cdot & \cdot \\ k_{m,1} & \cdot & \cdot & \cdot & k_{m,m} \end{bmatrix} \begin{bmatrix} x_1 \\ \cdot \\ x_n \\ \cdot \\ x_n \end{bmatrix} = \begin{bmatrix} a_1 \\ \cdot \\ f \\ \cdot \\ a_m \end{bmatrix}$$

Where f is the value that the nth degree of freedom must be taken.

Applications

Solid mechanics is usually modelled through a displacement-based model, thus Dirichlet boundary conditions usually consist in imposing the displacement of the structure in given points. Structural mechanics is often based on formulations which include relative rotations, whose numerical resolution requires nonlinear shape functions in the finite element approximation. For instance, frame structures are based on the beam theory and the related finite element has 6 degrees of freedom (3 displacements + 3 rotations in a 3D space). For a 2D problem, each node of the boundary has 3 degrees of freedom on which Dirichlet boundary conditions can be applied: 2 displacements (u_x and u_y) and 1 rotation (ω). These constraints are usually depicted as follows:

Symbol	Constraints
	$u_x = 0$ $u_y = 0$ $\omega = 0$
	$u_x = 0$ $u_y = 0$
	$u_y = 0$
	$u_y = 0$ $\omega = 0$
	$\omega = 0$

Table: External constraints to the ground

In the table, external constraints to the ground (i.e. set the value to zero) are reported, but the symbols are used also to express a fixed displacement/rotation different than zero.

Mixed Boundary Condition

In mathematics, a mixed boundary condition for a partial differential equation defines a boundary value problem in which the solution of the given equation is required to satisfy different boundary conditions on disjoint parts of the boundary of the domain where the condition is stated. Precisely, in a mixed boundary value problem, the solution is required to satisfy a Dirichlet or a Neumann boundary condition in a mutually exclusive way on disjoint parts of the boundary.

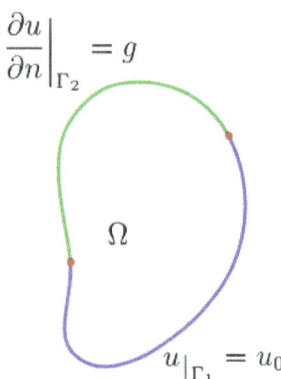

Green: Neumann boundary condition; purple: Dirichlet boundary condition.

For example, given a solution u to a partial differential equation on a domain Ω with boundary $\partial\Omega$, it is said to satisfy a mixed boundary condition if, consisting $\partial\Omega$ of two disjoint parts Γ_1 and Γ_2, such that $\partial\Omega = \Gamma_1 \cup \Gamma_2$, u verifies the following equations:

$$u\big|_{\Gamma_1} = u_0 \text{ and } \frac{\partial u}{\partial n}\bigg|_{\Gamma_2} = g,$$

where u_0 and g are given functions defined on those portions of the boundary.

The mixed boundary condition differs from the Robin boundary condition in that the latter requires a linear combination, possibly with pointwise variable coefficients, of the Dirichlet and the Neumann boundary value conditions to be satisfied on the whole boundary of a given domain.

Stefan Condition

A condition describing the law of motion of the boundary between two different phases of matter and expressed as a law of energy conservation under phase transformation. For example, the boundary between the solid and liquid phases of matter in a solidifying process (or a melting process) can be described in the one-dimensional case by a function $\xi = \xi(t)$, connected to the temperature distribution $u(x,t)$ by means of the Stefan condition:

$$\lambda\rho_1 \frac{d\xi}{dt} = k_1 \frac{\partial u(\xi(t) - 0, t)}{\partial x} - k_2 \frac{\partial u(\xi(t) + 0, t)}{\partial x}, \qquad t > 0.$$

The mass solidifies (or melts) in the course of time Δt.

$$\rho_1 \Delta\xi = \rho_1[\xi(t + \Delta t) - \xi(t)]$$

The amount of heat $\lambda\rho\,\Delta\xi$ thus required is equal to the difference between the amounts of heat passing through the boundaries $\xi(t)$ and $\xi(t + \Delta t)$:

$$\lambda\rho_1 \Delta\xi = \left[k_1 \frac{\partial u(\xi(t) - 0, t)}{\partial t} - k_2 \frac{\partial u(\xi(t + \Delta t) + 0, t + \Delta t)}{\partial x} \right] \Delta t.$$

Hence, when $\ddot{A}t \to 0$, the Stefan condition is obtained. Moreover, the temperature on the boundary between the two phases $\xi = \xi(t)$ is assumed to be continuous and its value is taken equal to the known temperature of melting.

Similar conditions on unknown boundaries which arise in studies on certain other processes and which follow from conservation laws are also called Stefan conditions.

Robin Boundary Condition

In mathematics, the Robin boundary condition, or third type boundary condition, is a type of boundary condition, named after Victor Gustave Robin .When imposed on an ordinary or a partial differential equation, it is a specification of a linear combination of the values of a function and the values of its derivative on the boundary of the domain.

Robin boundary conditions are a weighted combination of Dirichlet boundary conditions and Neumann boundary conditions. This contrasts to mixed boundary conditions, which are boundary conditions of different types specified on different subsets of the boundary. Robin boundary conditions are also called impedance boundary conditions, from their application in electromagnetic problems, or convective boundary conditions, from their application in heat transfer problems.

If Ω is the domain on which the given equation is to be solved and $\partial\Omega$ denotes its boundary, the Robin boundary condition is:

$$au + b\frac{\partial u}{\partial n} = g \qquad \text{on } \partial\Omega$$

for some non-zero constants a and b and a given function g defined on $\partial\Omega$. Here, u is the unknown solution defined on Ω and $\partial u / \partial n$ denotes the normal derivative at the boundary. More generally, a and b are allowed to be (given) functions, rather than constants.

In one dimension, if, for example, $\Omega = [0,1]$, the Robin boundary condition becomes the conditions:

$$au(0) - bu'(0) = g(0)$$

$$au(1) + bu'(1) = g(1).$$

Notice the change of sign in front of the term involving a derivative: that is because the normal to $[0,1]$ at 0 points in the negative direction, while at 1 it points in the positive direction.

Application

Robin boundary conditions are commonly used in solving Sturm–Liouville problems which appear in many contexts in science and engineering.

In addition, the Robin boundary condition is a general form of the insulating boundary condition for convection–diffusion equations. Here, the convective and diffusive fluxes at the boundary sum to zero:

$$u_x(0)c(0) - D\frac{\partial c(0)}{\partial x} = 0$$

Where D is the diffusive constant, u is the convective velocity at the boundary and c is the concentration. The second term is a result of Fick's law of diffusion.

Cauchy Boundary Condition

In mathematics, a Cauchy boundary conditions augments an ordinary differential equation or a partial differential equation with conditions that the solution must satisfy on the boundary; ideally so to ensure that a unique solution exists. A Cauchy boundary condition specifies both the function value and normal derivative on the boundary of the domain. This corresponds to imposing both a Dirichlet and a Neumann boundary conditions. It is named after the prolific 19th-century French mathematical analyst Augustin Louis Cauchy.

Second-order Ordinary Differential Equations

Cauchy boundary conditions are simple and common in second-order ordinary differential equations,

$$y''(s) = f\big(y(s), y'(s), s\big),$$

where, in order to ensure that a unique solution $y(s)$ exists, one may specify the value of the function y and the value of the derivative y' at a given point $s = a$, i.e.,

$$y(a) = \alpha,$$

and

$$y'(a) = \beta,$$

where a is a boundary or initial point. Since the parameter s is usually time, Cauchy conditions can also be called *initial value conditions* or *initial value data* or simply *Cauchy data*. An example of such a situation is Newton's laws of motion, where the acceleration y'' depends on position y, velocity y', and the time s; here, Cauchy data corresponds to knowing the initial position and velocity.

Partial Differential Equations

For partial differential equations, Cauchy boundary conditions specify both the function and the normal derivative on the boundary. To make things simple and concrete, consider a second-order differential equation in the plane

$$A(x,y)\psi_{xx} + B(x,y)\psi_{xy} + C(x,y)\psi_{yy} = F(x,y,\psi,\psi_x,\psi_y),$$

Where $\psi(x,y)$ is the unknown solution, ψ_x denotes derivative of ψ with respect to x etc. The functions A,B,C,F specify the problem.

We now seek a ψ that satisfies the partial differential equation in a domain Ω, which is a subset of the xy plane, and such that the Cauchy boundary conditions

$$\psi(x,y) = \alpha(x,y), \quad \mathbf{n} \cdot \nabla \psi = \beta(x,y)$$

hold for all boundary points $(x,y) \in \partial\Omega$. Here $\mathbf{n} \cdot \nabla\psi$ is the derivative in the direction of the normal to the boundary. The functions α and β are the Cauchy data.

Notice the difference between a Cauchy boundary condition and a Robin boundary condition. In the former, we specify both the function and the normal derivative. In the latter, we specify a weighted average of the two.

We would like boundary conditions to ensure that exactly one (unique) solutions exist, but for second-order partial differential equations, it is not as simple to guarantee existence and uniqueness, as it is for ordinary differential equations. Cauchy data are most immediately relevant for hyperbolic problems (for example, the wave equation) on open domains (for example, the half plane).

Neumann Boundary Condition

The Neumann boundary condition is a type of boundary condition, named after Carl Neumann given in below Figure. When imposed on an ordinary or a partial differential equation, it specifies the values that the derivative of a solution is going to take on the boundary of the domain. Given, for example, the Laplace equation, the boundary value problem with the Dirichlet b.c. is written as:

$$\Delta\varphi(\underline{x}) = 0 \qquad \forall \underline{x} \in \Omega$$

$$\frac{\partial\varphi(\underline{x})}{\partial n} = f(\underline{x}) \qquad \forall \underline{x} \in \partial\Omega$$

where n is the unit normal to the boundary surface, if $\Omega \subset R^3$.

Carl Neumann

In the case of ordinary differential equations (i.e. $\Omega \subset R^1$), the derivative normal to the boundary coincides with the global derivative φ'. In the rare cases in which a temporal dependency is solved through a finite element approach (instead of the more usual finite difference), this type of boundary condition is the most common. Neumann boundary condition is also called "natural" because it naturally appears in the development of the weak formulation in any finite element approach. Let's consider the following simple equation:

$$-u''(x) = p(x) \qquad \forall x \in \mathbb{R}$$

where u is an unknown scalar field and p is a given scalar function. This equation rules many phenomena, for instance, the thermal diffusion in 1D and the tension/compression of a beam. The finite element method consists in rewriting the equation from a differential (strong) form to an integral (weak) formulation. This transformation is done through two steps:

1. Test and integrate:

$$-\int_a^b u''(x)v(x)\,dx = \int_a^b p(x)v(x)\,dx$$

where $v(x)$ is the shape function.

2. Apply the Green theorem to have a uniform distribution of derivatives and avoid higher order derivatives:

$$-\int_a^b u'(x)v'(x)\,dx = \int_a^b p(x)v(x)\,dx + \left[u'(x)v(x)\right]_a^b$$

Thus, a term including the derivative of the unknown field on the boundary naturally appears; for 1D problems, this term refers to the extremes of the interval, for 2D problems it refers to the contour of the domain, and for 3D problems it refers to the boundary surfaces. The presence of the boundary term on the right-hand side highlights two properties of Neumann boundary conditions:

- Homogeneous Neumann b.c. are naturally satisfied without any explicit imposition

- Since Dirichlet b.c. are usually applied by modifying the right-hand term, a homogeneous Neumann condition is applied on all the boundaries where a Dirichlet b.c. is imposed.

Application

In solid mechanics, the spatial derivatives of displacements are related to the strain tensor. In elasticity, the strain is proportional to the stress; hence the Neumann boundary condition refers to both imposed strains and stresses. Since the stress is also linked to the external forces through the Cauchy's stress principle, the Neumann condition is also used to apply external loads. As stated in the section dedicated to Neumann boundary conditions, the homogeneous condition is naturally satisfied, so "free" boundaries may not be modelled explicitly.

References

- Guru, Bhag S.; Hızıroğlu, Hüseyin R. (2004), Electromagnetic field theory fundamentals (2nd ed.), Cambridge, UK – New York: Cambridge University Press, p. 593, ISBN 0-521-83016-8.

- Boundary-value-problem: scholarpedia.org, Retrieved 18 April 2018

- Dirichlet-problem, science: britannica.com, Retrieved 28 June 2018

- J. E. Akin (2005). Finite Element Analysis with Error Estimators: An Introduction to the FEM and Adaptive Error Analysis for Engineering Students. Butterworth-Heinemann. p. 69. ISBN 9780080472751.

- What-are-boundary-conditions, numeric: simscale.com, Retrieved 16 May 2018

- Stefan-condition: encyclopediaofmath.org, Retrieved 31 March 2018

- What-are-boundary-conditions: simscale.com, Retrieved 19 April 2018

Permissions

Index

www.ingramcontent.com/pod-product-compliance
Lightning Source LLC
Chambersburg PA
CBHW080406190526
45161CB00003B/154